Beyond Market and Hierarchy

Beyond Market and Hierarchy

Patriotic Capitalism and the Jiuda Salt Refinery, 1914–1953

Kwan Man Bun

BEYOND MARKET AND HIERARCHY
Copyright © Kwan Man Bun, 2014.

All rights reserved.

First published in 2014 by
PALGRAVE MACMILLAN®
in the United States—a division of St. Martin's Press LLC,
175 Fifth Avenue, New York, NY 10010.

Where this book is distributed in the UK, Europe and the rest of the world, this is by Palgrave Macmillan, a division of Macmillan Publishers Limited, registered in England, company number 785998, of Houndmills, Basingstoke, Hampshire RG21 6XS.

Palgrave Macmillan is the global academic imprint of the above companies and has companies and representatives throughout the world.

Palgrave® and Macmillan® are registered trademarks in the United States, the United Kingdom, Europe and other countries.

ISBN: 978–1–137–33526–5

Library of Congress Cataloging-in-Publication Data

Kwan, Man Bun, 1955–
  Beyond market and hierarchy : patriotic capitalism and the Jiuda salt refinery, 1914–1953 / Kwan Man Bun.
    pages cm
  Includes bibliographical references and index.
  ISBN 978–1–137–33526–5 (hardback : alk. paper)
  1. Jiu da jing yan gong si—History. 2. Salt industry and trade—China—History. 3. Government monopolies—China—History. 4. Capitalism—China—History. 5. Salt—Taxation—China—History. 6. China—History—20th century. I. Title.

HD9213.C44J585 2014
338.7'66440951—dc23   2014000894

A catalogue record of the book is available from the British Library.

Design by Newgen Knowledge Works (P) Ltd., Chennai, India.

First edition: July 2014

To Mother, Liz, and Tim

Contents

List of Tables	ix
Acknowledgments	xi
List of Abbreviations	xiii
Introduction	1
1 The Institutions	9
2 The Networks	25
3 Breaking into the Market	43
4 The Price of Success	61
5 Cartel as a Business Network	77
6 Relationship with the Nationalist State	95
7 At War	115
Postscript	135
Appendices	141
Notes	171
Glossary	217
Selected Bibliography	221
Index	231

Tables

1.1	Salt revenue in national budget: Regular income, 1912–1916	20
2.1	Identified and unidentified shareholders lists, 1915	27
2.2	Liang Qichao and his network in Jiuda, early December 1915	28
2.3	Chemical analysis of Changlu crude salt	40
2.4	Cost analysis of refined salt (NaCl at 98%)	40
4.1	Extractions from Jiuda by the Hunan provincial government, 1921–1925	68
4.2	Jiuda investments (book value) in Yongli and Yongyu, 1921–1937	74
4.3	Bank loans Jiuda diverted to Yongli and Yongyu, 1922–1937	74
4.4	Loan from Jincheng Bank to Yongli (with secondary guarantee by Jiuda)	75
5.1	Crude salt versus refined salt shipment: Four Yangzi provinces, 1931–1936	88
5.2	Cost comparison of refined salt and crude salt, Hankou 1933	89
6.1	NSRB calculations of registration allotment	106
6.2	State-sanctioned registration/market share, June 5, 1933	107
6.3	Sales of refined salt in Hankou, 1932–1936	108
7.1	State salt collection and distribution, 1942–1944	125
7.2	Salt-derived revenue for the Nationalist government, 1937–1945	125

Tables

Acknowledgments

Many friends and colleagues made this book possible. One hot summer afternoon two decades ago, following up on Professor Teng Weizhao's introduction, I called on Professor Guo Shihao, doyen of modern Chinese economic history at the Institute of Economics, Nankai University. Overcoming my reluctance, he insisted on receiving me in the garden of the Tianjin Municipal Hospital where he was awaiting surgery. As Professor Guo shared enthusiastically everything he knew about Jiuda and Yongli, location of various archival sources and the major issues to be explored, what I intended to be a brief visit stretched into hours. His passion rekindled my interest in the subject, and we agreed to collaborate on our research. That was the last time I saw him. The collaboration, however, continued with his students Professors Zhao Jin, Gu Yun, and Li Jianying, resulting in a five-volume set of selected archival sources on the conglomerate. Also at Nankai University, Professor Liu Foding made available archival materials left unpublished from the five volume set on modern China's salt administration, which he helped edit. Generous grants from ACLS, the Chiang Ching-kuo Foundation, and the Charles Phelps Taft Memorial Fund financed research trips to China, England, and Taiwan. The anonymous reader for Palgrave Macmillan provided an insightful critique and suggested many helpful ways to improve the manuscript. Judith Daniels suffered through my insufferable prose to make the book readable. My wife Liz and son Tim tolerated my disappearance into the basement and frequent absence from home to visit the various archives. I dedicate this book to my mother. Ever patient waiting for my infrequent and brief visits, she always insists that my work comes first. All errors, of course, are mine alone.

Abbreviations

AH	Academia Historica, Xindian, Taibei
AS	Institute of Modern History Archives, Academia Sinica, Taibei
CLHB	Archives of the Changli Division held at the Hebei Provincial Archives, Shijiazhuang
DGB	*l'Impartial*, Tianjin and Chongqing
FO	Foreign Office Archives, National Archives, Kew, U.K.
HSBC	Archives of the Hongkong and Shanghai Banking Corporation, Ltd., London, U.K.
JD	Unpublished materials from the archives of Jiuda Salt Refinery held at the College of Economics, Nankai University
JDJYZJ	Zhao Jin, Kwan Man Bun, and Gu Yun, eds. *Fan Xudong qiye jituan lishi ziliao huibian: Jiuda jingyan gongsi zhuanji* [Selected Archival Material of Fan Xudong's Industrial Group: Jiuda Salt Refinery]. 2 vols. Tianjin: Tianjin renmin chubanshe, 2006. Reprinted 2010
JDSYJ	*Jindaishi yanjiu* [Studies in Modern History]
JY	Unpublished materials from the National Salt Refinery Association held at the College of Economics, Nankai University
SM	Shanghai Municipal Archives
TJWSZL	*Tianjin wenshi ziliao xuanji* [Selected Materials on Tianjin History and Literature]
TM	Tianjin Municipal Archives
YL	Unpublished materials from the Yongli Soda Ash Manufacturing Company held at the College of Economics, Nankai University
YLZJ	Zhao Jin, Kwan Man Bun, Gu Yun, and Li Jianying, eds. *Yongli huaxue gongye gonshi zhuanji* [Yongli Chemical Industries]. 3 vols. Tianjin: Tianjin remin chubanshe, 2010

YMZK	*Yanmi zhuankan* [Saltophile Journal]. Beijing: Yanzheng zazhishe, 1935
YYSYJ	*Yanyeshi yanjiu* [Studies on the History of Salt Industry]
YZCK	*Yanzheng congkan* [Anthology on Salt Administration]. Beijing: Yanzheng zazhishe, 1921, 1931
YZZZ	*Yanzheng zazhi* [Journal on Salt Administration]
ZGJDYWSZL	Nankai daxue jingji yanjiusuo jingjishi yanjiushi, comp. *Zhongguo jindai yanwushi ziliao xuanji* [Selected Materials on Salt Affairs in Modern China]. 4 vols. Tianjin: Nankai daxue, 1985
ZGJDYWSZLBY	Unpublished rescripts from the Archives of the Salt Administration held at the College of Economics, Nankai University
ZGJJSYJ	*Zhongguo jingjishi yanjiu* [Studies on Chinese Economic History]
ZGSHJJSYJ	*Zhongguo shehui jingshi yanjiu* [Studies on Social and Economic History of China]
ZGYZSL	Caizhengbu Yanwushu, comp. *Zhongguo yanzheng shilu* [A Veritable Record of China's Salt Administration]. First compilation. 4 vols. Nanjing: Caizhengbu, 1933. Second compilation. 2 vols. N.p.: Caizhengbu, 1943

Introduction

Reviewing his work three decades later, Fan Xudong (1883–1945) declared that Jiuda, as the first-born of his industrial group, is truly a "Chinese-style" elder brother.[1] Indeed, from 1914 until 1953 when it became a part of the joint state–private Yongli-Jiuda Chemical Industries, Jiuda Salt Refinery Co., Ltd. was synonymous with modern China's refined salt industry. Building this private enterprise from scratch to national prominence, Fan and his colleagues promoted the country's health championing the consumption of hygienic refined salt, self-sufficiency in various chemicals, and industrialization with an "almost religious...fervour."[2] Beginning with a nominal capital of 50,000 yuan, by 1924 the company boasted a paid-up capital of 2.1 million yuan. With plants at Tanggu (Hebei), Dapu (Zhejiang), and the Yongyu Salt Refinery Co., Ltd. in Qingdao, Shandong, Jiuda controlled nearly half of the country's refined salt production capacity by 1937. Through Hengfengtang, its holding corporation, the company owned over 100,000 *mu* (one *mu* approximately one-sixth acre) of salt ponds, Tanggu's central mart and wharf, as well as partnerships distributing both refined and crude salt. Jiuda's employees were socialized, whether as a family or a corporate unit (*danwei*) through a variety of institutions and measures. Beginning in 1920, the company subsidized a clinic, mess hall, dormitory, tennis courts, and a savings scheme offering 10 percent monthly interest, followed in 1921 by a workers' cooperative, reading room, and literacy classes. For research and development, Fan and his colleagues endowed the Golden Sea Chemical and Industrial Research Institute (Huanghai huaxue gongye yanjiushe) in 1922. Construction of an imposing corporate headquarters in the heart of Tianjin's French Concession began in 1923. Star (Mingxing) kindergarten and primary school with free tuition followed in 1925. Over a thousand workers and their families proudly called Tanggu their home, transforming the sleepy seaside fishing village into a vibrant company town served by Jiuda's power plant.³

With Fan's determined support and the company's capital, insider loans, guarantees, and credit line, Yongli Soda Ash Manufacturing Co. (Pacific Alkali Co., Ltd.) was launched in 1917 and took ten years to unravel the Solvay process, finally dividing the country's soda ash market with Brunner, Mond Co., Ltd. (a division of Imperial Chemical Industries after 1926) through a market-sharing and price-fixing agreement. Risking everything yet again in 1933, Fan volunteered his group's capital and expertise to build an ammonia sulfate plant, breaking China's nitrogen bottleneck. Before the full-scale Japanese invasion in 1937, Fan controlled assets of 30 million yuan, although he owned neither a house nor an automobile, and for many years drew a salary only from Jiuda. Whereas profit maximization and pursuit of private interest led to greatest social good, at least in the Smithian tradition, Fan defined his patriotic capitalism as "no personal sacrifice was too great for the benefit of the country."[4] His many industrial schemes were designed to strengthen the country, to fulfill one's duty as a Chinese, and not for profit alone, much less personal gain.[5]

Fan's quest to chart a third path of national development between private and state enterprise thus intersects with three interrelated debates in modern Chinese history: the political economy of state and nation-building, the relationship between private capital (and capitalists) and the government, and social networks as a form of capital and business organization.

Since the late Qing, generations of statesmen, scholars, and entrepreneurs have wrestled with the problem of how to make China strong and prosperous again. Condemned to modernize amid foreign imperialism, indemnities to pay, and domestic strife, with private capital unwilling or unable to invest, government-led industrialization through adopting the latest scientific and technological advances, nurturing a healthy citizenry, and nation-building by a strong centralized state became part of the controversial praxis of public finance and industrial policy.[6] The role of private enterprise in industrial development remained ambivalent into the republican period. Sun Yat-sen's "Principle of the Three Peoples" called for a controlled economy (*tongzhi jingji*) and limited involvement of private capital in industries deemed vital to the nation, leaving the role of private enterprise in question.[7] Jiang Jieshi followed Sun's direction for state ownership of large industrial enterprises to prevent wealth disparity by restricting the growth of private capital (*jiezhi ziben*).⁸ The problem, of course, was how to raise enough revenue to meet his many needs.

As one of the major sources of revenue, the gabelle, a tax on salt "always hated everywhere" thus occupied a central place in the republican discourse on public finance.⁹ Should salt be free of tax and its production and marketing open to anyone? If it must be taxed, should the burden be equitable spatially, that is, uniform all over the country, regardless of income (however defined), ease of supply, varying production cost and ability to evade, or progressive according to income or distance from production points? From an administrative perspective, efficiency could be achieved in terms of collecting the tax directly from consumers or indirectly through agents, upon production or distribution, revenue maximization or stability, and the cost of administering the system. Simplicity as a desirable goal also affects compliance and perceptions of equity. A simple system might not generate enough revenue or be equitable, although a complex system designed for equitability could add to the burden and cost of collection.¹⁰ How, then, might the people's welfare (*minshi*), hygiene, and health be ensured while satisfying revenue needs to forge a strong and prosperous nation?

The Revenue Inspectorate, a multinational bureaucracy established under the terms of the Reorganization Loan of 1913, added to the debate. While contemporaries dismissed it as another attempt by foreign imperialists to control China, scholars have argued that this "synarchy"—efficient, professional, insulated, and meritocratic—successfully stabilized and modernized the country's finances by creating a system for the collection and auditing of salt taxes.¹¹ While revenue continued to flow into government coffers through the foreign banks servicing the loan, other scholars also have argued that the goals of modernization and revenue maximization are fundamentally incompatible with each other.¹² To the extent the Revenue Inspectorate and Jiuda shared the common goals of free marketing, lower price, and better quality of salt, the company's experience underscores the contradictions between revenue needs and the modern state-building during the republican period.

That experience is also part of the evolving relationship between government and private capital during the period. In communist historiography, the Nationalists owed their rise to collusion with the so-called Jiangsu–Zhejiang capitalists. On the other hand, Lloyd Eastman, Parks Coble, and others have argued that the republican state exploited the capitalists ruthlessly in pursuit of its own interests.

Still others characterized that relationship as shifting, ad hoc in nature, and difficult to generalize.[13] Missing in this debate is the critical role of the revenue farmers whose ability to advance the gabelle underwrote many of the early loans sustaining the young Nationalist government. To perpetuate their monopolistic powers in a "generally unrestricted competitive" economy, they bargained with Nanjing and paid for renewal of their hereditary privilege.[14] While the salt refineries and the Revenue Inspectorate shared the goals of free marketing and improved quality of salt, the need for revenue certainty thus divided them. Collusion or exploitation does not exhaust the complex patterns of interaction between the republican state and the economy.

Beyond the political economy of the gabelle, how Fan Xudong and his colleagues managed Jiuda also challenges our understanding of the firm, its operation, and evolution in modern China. A paradigm of failure, nepotism, cronyism, and paternalism, characteristic of traditional Chinese businesses, marred economic development during the period, the golden age of the Chinese bourgeoisie from 1927 to 1937 notwithstanding.[15] As merchants began to assert their collective interests through institutions such as chambers of commerce, the resulting network revolution reinforced an oversocialized, if not a "trendy" enigmatic Chinese capitalism and business culture characterized by particularistic ties, collusion, and corruption.[16] On the other hand, quite apart from the ideotypical, if not false, dichotomy of tradition and modern, recent studies on family business and hybrid forms of business organization have revealed a far more nuanced picture. Paternalism did not prevent some Chinese entrepreneurs from consulting and delegating authority.[17] Whether because of a theoretical bias or a lack of empirical evidence, or both, Asian network remains an orientalist enigma endowed with mythical powers even though similar business networks could be identified in Europe and the Americas.[18]

Drawing from personal correspondence, company and government papers, this book continues the dialog between general theory and the particular reality of modern China by focusing on the complementarity and contradictions of dominant institutions (including cultural, social, financial, and legal systems) and their ways of coordination and control of economic activities among the firm, market, and the state in modern China's salt industry.[19] Specifically, the analysis explores four interlinked processes: the role of political power and institutions in shaping the refined salt business; Jiuda's social embeddedness as an economic actor; its ownership, coordination, and governance systems as it grew; and its complex relationships with other refineries, revenue farmers, and the republican state.

Chapter 1 analyses the institutional context of the salt trade that shaped Jiuda and the refined salt industry. While Chinese reformers debated about what to do, the Reorganization Loan promised the modernization of the salt industry, including the establishment of a state-owned mechanized salt refinery. Archival sources, however, reveal a fractious international bureaucracy beset by competing national interests, conflicts over revenue security, and avowed goals of modernizing the industry. To achieve the mission they set for themselves, Fan Xudong and his colleagues had to learn how to exploit these conflicts.

Chapter 2 analyses the embedded structure, operation, and resources of Jiuda's spatially and hierarchically differentiated shareholder network. Beyond the family, weak ties based on common native place, shared goals and life experiences, including school and study abroad, gave it strength. Key shareholders served as structural holes for recruitment and network cohesion. In addition to furnishing monetary capital, Jiuda's shareholder network also mobilized resources to overcome bureaucratic hurdles and gain access to the salt market.

A war of salts followed, the subject of chapter 3. As Jiuda commodified a modern and hygienic product to discerning consumers, crude salt producers, and revenue farmers, not to mention bureaucrats with vested interests in the division system, fought the upstart. Battle was not restricted merely to the marketplace but extended to the political and legal systems. The rule of law might be found wanting and redress seldom sought through the law in modern China, but the company and the revenue farmers' wranglings with Beijing's Administrative Court, the Salt Administration, and the Revenue Inspectorate, at times Kafkaesque, captures how political and legal institutions shaped this sector of the republican economy.

Chapter 4 highlights the price Jiuda paid for its success as well as the constraints of its network strategy. Moving beyond a model of business organization progressing lineally from tradition to modern, Jiuda evolved as a hybrid, if not cross-vergent, company combining both Western and Chinese business practices.[20] Limited liability, board of directors, regular shareholders' meetings, annual reports, publicly traded shares, floating of company bonds, centralized hierarchical management, codified governance, double-entry accounting, eight-hour work day, generous worker benefits, and other practices such as declaring dividends on net profit (instead of fixed dividend, or *guanli*) coexisted with unlimited liability partnerships, holding corporations, labor bosses, and networks based on a variety of attributes. On the other hand, the networking strategy, certainly not without maintenance costs, limitations, and risks of its own, also exposed the company to the dark sides of social networking.

In dealing with market failures—overproduction, limited access, and unnecessary, if not wasteful, competition—Jiuda followed another network strategy by cartelizing the refined salt industry. As chapter 5 demonstrates, however, the line between collusion and competition is often murky, nor does cartel necessarily translate into supranormal profit.[21] Periodically negotiated joint sales agencies, market-sharing and price-fixing agreements helped stabilize the industry for possible joint industrial (as opposed to supranormal) profit to survive a difficult time. As the salt refineries learned to monitor output and price while designing elaborate rules against cheating, they also waged short-lived bargaining price wars and squabbled over enforcement and compliance issues even as they collaborated in lobbying officials to change gabelle policy.

Addressing the relationship between big business and the republican state, chapter 6 reexamines the complex interaction between the salt industry and the Nationalist government. Golden age for the Chinese bourgeoisie or not, Nanjing's growing revenue needs, rent creation, and extraction by officials underscored that interaction. In a tightly regulated industry such as refined salt, private capitalism became highly politicized as Fan and his colleagues dedicated much of their energy to dealing with the state and its officials. Behind a veil of official memos, orders, and a modernized salt administration, Jiuda's archive reveals how the salt refineries waged a war of contributions, manufactured relationships, and public opinion to shape the New Salt Law. It was a hollow victory, however, as fiscal realities and strategic concerns caused Nanjing to delay implementing the new law that promised to end the revenue farming system.

Indeed, Fan Xudong's practice of patriotic capitalism did not translate easily into profits, a subject explored in chapter 7. While he ignored complaints from his shareholders about dwindling dividends as he diverted resources to other industrial projects in the service of the country, the Second World War presented even greater challenges to Jiuda's survival. Among the few companies from Tianjin that followed the Nationalist government to the interior, Jiuda struggled to increase salt production in Sichuan to help make up for the loss of the coastal production fields and industrialization of the Southwest. With the nationalization of the salt industry in 1942, Jiuda became primarily a contract supplier to the state as the industry became an appendage to state banks and large private banks, foreboding the transformation of private industry to state-owned enterprise and planned economy after 1949.

This book is thus a microhistory of modern China's salt industry, revealing processes hitherto hidden from theoretical and historical accounts. It tells the story of Fan Xudong as a Schumpeterian entrepreneur using Jiuda to remake, if not destroy, institutions and forge new ones. Straddling the public and private, as well as the state and the economy, salt played a variety of roles—as an alimental necessity, a critical revenue source, as a business, and as an instrument for capital formation—in private, state and nation-building. But the process and the resulting enterprise were also a hybrid of the old and the new as Fan and his colleagues wrestled with the salt bureaucracy through networking and manufactured relationships.

As a case study of the embeddedness of economic behavior in culture and society, Jiuda's experience illustrates how family, norms, customary business practice, and heterarchical social networks were embedded in business, raised the capital it needed, managed its growth, and forged a system of norms to achieve a variety of goals (profit maximization being only one among several), which, in turn, reinforced these institutions (reciprocal embeddedness). Put another way, instead of being mutually exclusive of each other, market, hierarchy, and network coexist as three interacting modes of business behavior, coordination, and governance.[22]

The story of Jiuda is also a prosopography of Fan and his colleagues, a generation of capitalists who espoused a third path of national economic development. Unlike the Smithian capitalist, they intentionally pursued their calling not for personal gain or profit, but to champion modern science, hygiene, and China's development through a socially responsible private enterprise.[23] Fan's patriotism and his pursuit of the common good must be distinguished from official nationalism as mandated by the Nationalist state.[24]

1
The Institutions

As part of the state- and nation-building enterprise, the role of salt and its taxation generated considerable controversy in republican China. This chapter traces the evolving debate as Yuan Shikai (1859–1916, president 1912–1915) and his successors struggled to assert central control while the Reorganization Loan of 1913 secured on the gabelle created a seemingly "colonial and bizarre" Revenue Inspectorate charged with the mission of both collecting the tax to ensure repayment of the loans and modernizing the system on the principles of free marketing, equity, efficiency, and simplification. The Weberian ideotypical bureaucracy—professional, insulated, and departmentalized—seemingly delivered.[1] On the other hand, Esson Gale, among the inspectorate's earliest foreign staff, remained equivocal in explaining the apparent success: Yuan Shikai's military control of the provinces, inflation, or unknown revenue sources appearing "as if by magic."[2] Unraveling the secrets behind the gabelle growth, the competing interests of the central government, the provinces, officials, and warlords, as well as the foreign staff at the inspectorate (and the masters that they must serve) and resourceful revenue farmers all contributed to the continued morass that was salt (*yanhutu*). A steady flow of revenue became the guiding principle shaping not merely the institutions but also the salt business.

The debate

The abdication of the Qing dynasty renewed the debate. Dividing the country into salt divisions, with supply and distribution handled by hereditary revenue farmers, no longer seemed to serve the public good. The search for a solution drew inspiration from China's past and abroad. Yan Fu (1854–1921), echoing Adam Smith, argued that the regressive tax should be abolished altogether: already heaviest among nations, China's gabelle revenue benefited not the government but the unscrupulous revenue farmers.[3] To do away with all the corrupt merchants and bureaucrats, Tang Shouqian (1856–1917), echoing Feng Guifen (1809–1874), proposed that the gabelle should be folded into land tax, and salt allowed to be traded by everyone, as had been attempted in Gansu, Hedong, and parts of Changlu and Shandong divisions.[4] Even though it meant that the landed would be subsidizing the landless, this school argued that the property holders would surely accept such a modest burden as loyal citizens of the republic.[5]

Others did not share this charitable view. With all the differing rates within and among divisions, the regressive gabelle already represented an unfair burden on many people. The landed would simply increase the rent on their tenants if the gabelle became a part of the land tax. If there must be a tax on salt, then only the state should benefit from it, not the revenue farmers. Drawing on the Meiji experience, they argued that the state must standardize the tax for an equitable burden and assume direct control of salt production, collection, transportation, and distribution as part of the modern nation-building enterprise.[6] But where could the government find the capital, manpower, and expertise to implement such a system?

Before Beijing or Nanjing could decide, provincial authorities took the matter into their own hands. Hunan resolved to take over the salt trade, whereas Sichuan adopted free market.[7] In Lianghuai and elsewhere, local military commanders began putting the squeeze on the merchants. Soldiers seized the salt and began trading. Song Jiaoren (1882–1913), a founder of the Nationalist Party and a staunch critic of the revenue farming system, might welcome this de facto abolition of divisional boundaries, but other advocates of gabelle reform found themselves divided on what to do.

While sharing the same disdain for revenue farming and divisional boundaries, Zhang Jian (1853–1926) and others invoked the idea of Li Wen (1608–1647). The state simply did not have the expertise and the capital to take over the entire salt operation. Instead, it should concentrate on taxation at source (*jiuchang zhengshui*), and once the uniform gabelle was paid, anyone should be free to transport and sell salt anywhere. Smuggling would no longer exist, as there would be no more divisional boundaries. For a transitional period toward this ideal, Jing Benbai (1876–1948), founder and editor of the Society for the Discussion of Salt Administration (*Yanzheng taolunhui*), persuaded Zhang to modify his position somewhat: the state would maintain a monopoly over supply while specific marketing areas could still be designated and assigned to merchants under an arrangement known as *jiuchang zhuanmai*; they would, however, no longer enjoy any hereditary privilege. Zhang presented the proposal to the provisional national assembly in 1912.[8]

However, Zhang Jian soon found himself torn between meeting revenue needs while promoting reforms as president of Jing's society. Serving concurrently as the minister of industries and commissioner of the Lianghuai division, he had little choice but to implore revenue farmers to resume the trade and the flow of gabelle.[9] Xiong Xiling (1870–1937), another advocate of gabelle reform and vice-president of Jing's society, also found himself in a difficult situation. As the finance minister with a nearly empty national treasury, he needed, in addition to meeting the central government's payroll, seven million *taels* for the military each month as soldiers under both Yuan Shikai and Huang Xing threatened (and did) mutiny. In addition, 26 million *taels* in indemnity payments, foreign loans, and interest were due at the end of the year. The ministry once projected over 71 million yuan in salt revenue available to the central government, but his plea for the provinces to remit a share of the gabelle and other revenues, including the transit tax (*lijin*) on salt (*yanli*), due to the central government, went largely unheeded.[10] Of the total of 3.4 million *taels* pledged by ten provinces, only 400,000 *taels* found its way to the capital by May 1912, and 2.6 million *taels* by October 1913.[11] Strapped for cash, both Yuan Shikai and Sun Yat-sen agreed that revenue farming should continue, at least for the time being.[12] Worsening financial pressure on the young republic thus limited the options for any serious gabelle reform from the beginning.

Salt and the Loan Tangles, 1912

Both Yuan Shikai and Sun Yat-sen also agreed on another need: foreign loans. In late February 1912, the young republic began negotiations with the Four Power Groups Consortium of British, French, German, and American banks led by the Hongkong and Shanghai Bank Corporation. The result became known as the Reorganization Loan of 1913, a controversial episode condemned as another treasonous act by Yuan Shikai and as a violation of China's sovereignty by foreign economic imperialism.[13]

Neither generalization is entirely accurate. Yuan Shikai did broach the idea of a major foreign loan soon after he returned to Beijing as premier on November 13, 1911, but Sun Yat-sen had already approached the Hongkong Bank for financial support. Charles Addis, the bank's London manager, insisted over dinner that a proper government must be established first.[14] While Yuan Shikai undoubtedly needed money to reorganize the country, including eventual suppression of Nationalists in the south who challenged his authority, Nationalist forces were also using the gabelle as collateral to raise money for a variety of purposes, including armaments purchase.[15] Both sides wanted foreign support and financing.

As agents of economic imperialism, the foreign banks approached the negotiation table with mixed motivations. Profits and national prestige were at stake. Working through the bankers, foreign diplomats demanded that China must assume full responsibility for the obligations incurred by the *ancien regime*, damages during the fall of the dynasty, and make payments on indemnities and loans already in default. The diplomats also wanted a stable country for foreign commerce and investments and the opportunity to secure control of the salt gabelle.[16]

Yuan Shikai and his colleagues played the loan game reasonably well, given their weak hand. With the country increasingly divided, the central and local governments in disarray, and mounting financial problems, there was general agreement by everyone, including even Liang Qichao (1873–1929), about the need for foreign loans. But the issue was on what terms. Instead of caving in to foreign demands, Yuan and his colleagues pursued reforms initiated in late Qing period and, to the extent possible, exploited weaknesses of the groups consortium and Whitehall's failure to control London's impertinent bankers.

Following tortuous negotiations, the Finance Ministry proposed to create a salt administration. Under it would be a revenue inspectorate with a Chinese chief inspector and a foreign chief inspector selected from the staff of the Maritime Customs, supported by other foreign staff stationed at various salt production centers and "commercial ports" (*tongshang kou'an*) near them while leaving salt transportation, distribution, and anti-smuggling operations to the divisional commissioners.[17] The Chinese choice of J. F. Oiesen, an experienced Danish member of the Maritime Customs as the foreign chief inspector, ignited a diplomatic row. Denmark's neutrality was deemed inadequate for protecting the interests of the groups consortium.[18] When the dust finally settled, Sir Richard Morris Dane (1854–1940), retired Inspector General of Excise and Salt for India, became the foreign chief inspector, with other positions awarded to nationals of the groups consortium.

Although it could have been worse, the terms of the loan were draconian. Repayable over forty-seven years, the £25 million loan, 84 firm, carried a nominal interest rate of 5 percent (after discount and a commission of £1.5 million for the banks, an effective interest rate of approximately 6.4 percent, doubling London's interbank discount rate of 3 percent), and a handling charge calculated at ¼ percent of the annual loan service. Of the £21 million receivable, over half immediately went toward various overdue indemnities, repayment of advances, and provincial loans. Beijing received less than £10 million (or 90 million yuan), of which £8.5 million were earmarked for regular government expenditures and demobilization. There was no provision for direct foreign supervision of these disbursements except through the foreign staff employed by the Accounts and Audit Department, and the Bureau of National Loans, Ministry of Finance. Buried at the end of the agreement, Annex F set aside 20 million yuan (say £2 million), or 8 percent of the loan, for the reorganization of the salt industry: seven million as capital for state purchase and transportation of salt; three million for a salt factory; and five million each for reorganization of salt lands and advances to salt merchants.[19]

Politics and violation of parliamentary procedure aside, the offending provisions in the loan agreement that caused the most uproar read:[20]

> Article IV: The entire loan together with any advances...is secured in respect to both principal and interest by a charge upon the entire revenues of the Salt Administration of China...
>
> Article V: The Chinese Government engages to take immediate steps for the reorganization with the assistance of foreigners of the system of collection of the salt revenues of China...The Chinese government will establish a Central Salt Administration (*Yen wu shu*) at Beijing under the control of the Minister of Finance. This Central Salt Administration will comprise the Chief Inspectorate of Salt Revenues (*Chi ho Tsung So*) under a Chinese Chief Inspector (*tsung pan*) and a foreign Chief Inspector (*Hui pan*) who will constitute the chief authority for the superintendence of the issue of licenses and the compilation of reports and returns of revenues. In each salt-producing district there will be a branch office of the Chief Inspectorate (*Chi ho Fen So*), under one Chinese and one foreign District Inspector (*So chang*) who

shall be jointly responsible for the collection and the deposit of the salt revenues... Release of salt against payment of dues in any district will be made only under joint signature of the Chinese and foreign District Inspectors...

Critics of the agreement, agreeing with the U.S. president Woodrow Wilson, immediately identified loss of control over the gabelle as a major compromise of China's sovereignty.[21] Even worse, there was no term limit on foreign control of the "reformed" Salt Administration even after full repayment of the loan.

Carrying an above-market interest rate and attaching the entire salt revenue rather than the gabelle surplus, Article IV was necessary perhaps because of the risks of lending to an unsettled China and the "unproductive" use of the loan.[22] But the bankers (or the foreign governments behind them) did not stop there. Until the reformed Salt Administration could generate enough income for three successive years, revenue from Hebei, Shandong, Henan, and Jiangsu provinces would service the loan. In addition, after meeting all existing obligations, any surplus revenue of the Maritime Customs would also be hypothecated as a secondary guarantee. Indeed, a revamped Maritime Customs under Sir Francis Aglen, "purified and freed from the favoritism and nepotism which had obtained in the past (under Sir Robert Hart and others)," generated enough surplus to cover this loan service beginning in 1917.[23] Revenues from the Salt Administration were no longer needed, but the banks insisted on a 10 million yuan sinking fund in the gabelle account as reserve "sufficient... for a considerable period ahead" to service the loan, even though Dane protested that the withholding was illegal under the terms of the loan agreement.[24]

Nevertheless, the triple security to service the Reorganization Loan did not obviate the need for extraordinary foreign participation in reforming and supervising the Salt Administration, the clause "... so long as the interest and principal of this loan are regularly paid there shall be no interference with the Salt Administration..." notwithstanding.[25] Indeed, it was suggested that the Beijing government, specifically Zhou Xuexi (1866–1947), deliberately wanted foreign control. An official-industrialist with a reputation rivaling Zhang Jian's, he was no stranger to the salt business.[26] His family held substantial interests as a revenue farmer of the Lianghuai division. His critics (and political opponents) charged that Zhou surrendered the authority of issuing salt licenses (*yinpiao*, an official permit for salt in transit) to foreigners as part of the agreement so that his family's hereditary privilege would be preserved.[27]

In retaining the foreigners, Zhou and Yuan Shikai's motives might have been even more complicated. Their critics charged them with treason, but Zhou and Yuan might have been equally, if not more, interested in reasserting the central government's authority over provincial revenue. To create a fait accompli before the loan agreement was signed, Yuan had proposed a reformed Office of Salt Affairs (*Yanwushu*) to the provisional national assembly on November 10, 1912. In preparation for this transition, the office was put under the direct control of the Finance Ministry. Continuing the late Qing logic toward centralization, Yuan Shikai reiterated on January 15, 1913, that all salt revenue, regardless of its local or provincially designated use, must be reported, audited, and remitted to the new office.[28] By directing the flow of provincial salt revenue that the central government had known about since late Qing toward foreign banks under the pretext of repaying foreign loans, Yuan and Zhou were killing two birds with one stone: meeting revenue needs while using foreigners as agents to reassert Beijing's control over the provinces.[29] Even a staunch advocate of gabelle reforms, Liang Qichao conceded that, wounded pride aside, foreign imperialism, as articulated in the control of the inspectorate backed by gunboats, might have a better chance of collecting the revenue.[30]

Weak Institution and Polity

But how did the foreign staff deliver? On the one hand, the Salt Administration, or more accurately, the Revenue Inspectorate, has been lauded as a model of a modern and efficient organization in a chaotic time. At its best, it was a "synarchy," an organization insulated from Chinese politics and other particularistic influences, with a staff independently recruited, professionally trained, well-paid, and promoted on the basis of performance and experience.[31] On the other hand, Sir John Jordan, the British ambassador, remained skeptical; the idea that a handful of foreign advisers, auditors, and accountants could effectively manage the farflung salt administration was doubtful, if not illusory.[32] What did some "forty-odd" foreign staff—a quarter of them concentrated in Beijing, leaving a handful for most divisions—accomplish?

Upon arrival in Beijing, Dane felt that to reform China's salt administration, the fait accompli of the Office of Salt Affairs must be overturned. He and his foreign staff needed more authority than that granted explicitly by the loan agreement. With the support of the groups consortium and the legations, he successfully demanded the promulgation of more detailed regulations governing the workings of the chief and district inspectorates. The foreign chief inspector was equal in authority to the Chinese chief inspector and, as adviser to the Central Salt Administration, his expanded duties and authority is defined as follows:[33]

> Before any reorganization is effected of the existing system of salt administration and before any contract is concluded or any orders of importance are issued regarding the salt revenue and the transportation of salt to different localities for consumption (the official transportation, merchant's transportation and transportation by the public at large being included) the said adviser shall be consulted by the Head of the Central Salt Administration, and the nature of the decision to be arrived at and of the orders to be issued shall be determined by the Head of the Central Salt Administration in personal consultation with or in correspondence with the said adviser.

He must be consulted not merely on the collection and auditing of salt revenues for the central government, but also about all forms of salt transportation and supply in the entire country. Without his signature, no surplus gabelle revenue could be released to the Chinese.[34]

To Dane's credit as China's salt *imperium in imperio,* he did not replicate in China the British policy of government monopoly of Indian salt, closure of salt fields, erection of an impenetrable 2,500 mile-long customs barrier manned by 12,000 agents, with duties on native production equivalent to imports from England for the preservation of Cheshire exports.[35] Instead, the consummate technocrat and hard-working "salt gabbler" insisted on inspecting all the major divisions in person. He declared as his primary goals the building of a modern inspectorate, a uniform and comprehensive tax at source, and a large free trading area enjoying cheap salt of high quality.[36] Given these monumental tasks, his success was not unqualified.[37]

His problems began with the difficulties of recruiting a qualified staff. On the Chinese side, high-level appointments at the central and divisional administrations remained vulnerable to politics. Field staff training through a specialized school was not implemented until after 1920, but then its Chinese graduates were deemed unacceptable by the inspectorate.[38] Nor did Dane find all his foreign subordinates entirely up to his standards. Forced to recruit a multinational staff at short notice, he had had to compromise as "some appointees were scarcely fit to service as colleagues of Chinese officials of high rank and personal qualifications."[39] R. D. Wolcott, an American who joined the service in 1919 and rose to acting foreign chief inspector in 1941, observed with dismay:

> (Esson) Gale is right in that control in this Service is purely bureaucratic; there is no real attempt at justice or fair-play. China's civil services are just hunting-grounds for British, French, and Japanese adventurers.[40]

Nor was Dane insulated from power politics. He threatened resignation, though in vain, several times to protest against what he considered as bad examples of intervention by the British government, making him, an Irish, less than total as a loyal subject.[41]

Even as Dane labored to staff his far-flung inspectorate with foreign appointees, he needed to institute uniformity of tax rates and standardize currencies and weights before any serious audit, a task that was achieved with the assistance of his Chinese colleagues. In October 1913, Jing Benbai, Dane's counterpart as domestic adviser to the Salt Administration, drafted and promulgated new regulations for the standardization of weights and gabelle rate. Henceforth, each catty equaled 1.4 lbs., one *dan* of 100 catties weighed 140 lbs., and 16.8 *dan* made one ton. The country was divided into two demarcated by the Huai River. A national uniform rate of 2.5 yuan per *dan* was set, although until January 1915, a rate of two yuan applied north of the Huai, whereas south of the river the existing rate applied.[42] Some seven hundred categories of salt revenue were consolidated significantly, though still far from uniform. The various divisions continued to collect the gabelle in 67 grades while preserving the most glaring inequities inherited from the *ancien regime*: there was wide rate disparity among the divisions and, the farther away from the production field, the higher the tax. Ten years later, the number of grades had grown to seventy-four, and by 1929, eighty-three.[43] A frustrated Jing Benbai was thus highly critical of this increasing complexity and even more by the lack of transparency in the inspectorate's operation.[44]

Dane's record in promoting the free trade of salt was similarly equivocal. Drawing from his experience in India, he argued that replacement of the hereditary revenue farming system by free trade would result in an increase in salt sales and lowered prices, Zhou Xuexi's ploy notwithstanding. Before embarking on his inspection tour of the country, he recommended on December 11, 1913, that Changlu be used as a test case for "free" trade. Revenue farmers mounted a fierce lobbying campaign, and the resulting uproar forced Yuan Shikai to intervene. As a compromise, Zhang Hu (1875–1937), the Chinese chief inspector, declared the sixty-one districts seized by the state in 1911 and the thirteen districts of Zhangjiakou as free beginning July 1, 1914. State operation of the districts since 1912 had not been as profitable as once hoped. Under Zhang's reform, any duly registered salt dealer—the target was 3,000 to 3,500—would be allowed to participate in the trade. By September 14, 1914, 955 were on record. To attract more dealers, Dane reluctantly agreed to denominate the price of salt on the basis of the silver dollar instead of copper coinage, resulting in a de facto 30 percent price increase, a windfall for the dealers with a sizable inventory, and several million yuan of transactional donations (*yangma baoxiao*) to Yuan Shikai's coffers (but not to those of the groups consortium).[45]

Subsequent events and investigations hinted at how Zhang Hu undermined the free trade experiment. Despite Dane's vigorous defense of his colleague as a capable administrator and dedicated reformer, Zhang was compromised not merely by accepting 200,000 yuan from the Changlu revenue farmers for upholding their hereditary privilege.[46] He also went into business with wealthy revenue farmers and powerful generals and their associates to form the Changli Company to assume control of the freed districts and registration of dealers. The company, subsequently renamed and reorganized whenever political and military control of the provinces changed hands, continued as revenue farmer of the monopolies until the end of the Second World War.[47] Tianjin and Wuqing, together with the seven districts of Yongping prefecture, operated under similar arrangements during the period.[48] The reformed Salt Administration also did not stop local officials from collecting regular contributions from revenue farmers.[49] As the showcase of a modern gabelle administration under foreign control with free trade and lowered price of salt, Changlu fell short of Dane's own standards.

If his experiment to modernize Changlu was compromised by Zhang Hu's schemes, Dane's policies for the Lianghuai division also revealed the malleability of modernity and the rationality of revenue farming. Despite his refusal to legitimize the hereditary privilege of the Changlu revenue farmers, Dane came to a different conclusion after inspecting the Lianghuai division. In the four Yangzi provinces (Anhui, Hunan, Hubei, and Jiangxi), he encountered the best preserved hereditary revenue farming system in the country. Densely populated but with little production capacity and resource (except Yingcheng in Hubei), the provinces were ideally suited to generating a steady flow of revenue. He thus found that while the hereditary license holders enjoyed a healthy profit, it was only fair compensation for their service to the state. At any rate, "the government could not risk disturbing these revenues by precipitate reforms."[50] The major changes at Lianghuai effected under Dane's regime, after much negotiation between him and his Chinese counterparts, focused on the standardization of weights (in effect, another gabelle rate increase, see below) and the gabelle payment method. A nonrefundable 1.5 yuan gabelle must be paid before salt were released from the depots, with the remaining 3 yuan payable after the salt had been sold at the designated market, the 2.5 yuan per *dan* national average rate target notwithstanding. To prevent abuse, if any shipment was lost in transit, the merchant remained responsible for the entire gabelle due.[51] In Dane's view, a working machine needed no fixing, especially when its abolition would require compensation to the revenue farmers, an argument that Jing Benbai rejected most strenuously.[52]

Administration of the Lianghuai division, the largest and carrying the heaviest gabelle in the country, was thus left largely undisturbed by the modernization effort. In 1932, responding to charges of malpractices by salt merchants in the four Yangzi provinces, F. A. Cleveland, the foreign chief inspector, defended the integrity of the foreign staff:

> From 1913 until a recent date (ca. 1930), the authority of the Inspectorate to superintend the storage and distribution of salt, to access and collect revenue, and to release salt from stores or in transit after duty payment was contested by the Yenwushu as not extending to the four Yangtze provinces.... Political considerations determined the controversy in favor of the contention of the Yenwushu, thereby enabling the Transportation Officers to have full control....[53]

What the Revenue Inspectorate and the central government did have control over was the gabelle rate and the revenue farmers. Indeed, the magical appearance of revenue and its growth could be traced, in part, to changes in payment method and successive rises in the tax rate. Beginning with the late Qing reforms, the central government had sought accounting of some seven hundred categories of revenue derived from salt as part of the national budget. In 1913, all these "heterogeneous taxes, dues, and even 'voluntary contributions'" were consolidated into a single payment of 5.752 taels (say 8.628 yuan at 1 tael to 1.5 yuan) per *yin* of salt of 602 catties for Changlu.[54] Paid by the revenue farmers to the inspectorate, the division thus contributed 6.4 million yuan, much of which had hitherto been earmarked and withheld from the central government, or some 38 percent of the national salt revenue of over 16 million yuan remitted to Beijing, out of the budgeted 77 million yuan in 1913 (see Table 1.1).

Before the assistance of foreign inspectors began in May 1913, then, Beijing had taken steps to seize local and provincial salt revenue.[55] In addition, close to one-third of the revenue audited by Beijing for 1912 and 1913 came from price increases (see Table 1.1), part of the two *wen* per catty imposed by Xiong Xiling.

With Dane's approval, Jing Benbai also promoted gabelle rate uniformity as part of the reform. The first standardization of 1913 covering the country north of the Huai River resulted in a 45 percent increase from 1.38 yuan to 2 yuan per *dan* for Changlu, and the Northeastern provinces from 0.65 to 2 yuan, or a jump of over 200 percent, while small decreases were effected in several local markets temporarily.[56] However, for those divisions south of the Huai with a gabelle rate of at least 4.5 yuan per *dan*, the new "average national rate" of 2.5 yuan did not apply. The principles of "an old tax is a good tax" and "tax according to the inability to evade" still applied, nor was there any effort made to reduce the price of salt at the local retail level.

For Changlu, salt revenue after the establishment of the divisional revenue inspectorate in 1913 (auditing effective after December 1,1913) benefited from these reforms. The first phase of gabelle

Table 1.1 Salt revenue in national budget: Regular income, 1912–1916[a] (unit = yuan)

Year	Gabelle	Price Increases	Total Salt Revenue	% of Total Revenue
1912	25,078,317	25,976,366	71,279,902	20.3
1913	39,758,706	22,622,419	77,401,265	24.3
1914			84,879,873	24.2
1915			80,503,000[b]	34.03[c]
1916			84,771,365	20.1

Sources: [a] Jia Shiyi, *Minguo caizhengshi* [A Fiscal History of the Republic], 2 vols. (Shanghai: Shangwu yinshuguan, 1917), v. 2, 1580, 1600, 1615, 1630.
[b] *ZGJDYWSZL* IV: 266.
[c] Liu Cunliang (1934), 28.

standardization from January to September 1914 projected an additional revenue of 778,944 yuan. This was followed after September 1914 by yet another rate hike from 2 to 2.5 yuan per *dan,* yielding another 2.5 million yuan. For 1914, the first full year of the inspectorate's operation, Changlu reported a total revenue of over 12 million, an increase of almost 5 million from 1911, of which at least 3 million came from the rate standardization/increases.[57] One year later, Changlu's gabelle rate was raised by yet another 10 percent, or 0.25 yuan, to help pay for the 7 million taels loan due the Da Qing Bank, with total revenue reaching over 13 million.[58] Elsewhere, as a result of these gabelle rate standardizations (or increases), even larger gabelle yields were reported by the inspectorate for Shandong and the Northeast, the former held mostly by hereditary revenue farmers and the latter operated by the state.[59] Even the four provinces of Yangzi, already among those carrying the heaviest gabelle burden, were not spared increases, much to Sir Richard's regret.[60] By 1915, with the average national gabelle already in violation of the legal limit of 2.5 yuan per *dan,* the Salt Administration, with the concurrence of Dane, raised the limit to three yuan per *dan.*[61] By 1937, the national average gabelle had almost doubled to 5.98 yuan per *dan.*[62] All these rate increases, of course, only stimulated smuggling as the difference between production cost and after-tax price of salt widened. In a candid assessment of the inspectorate's efficacy, Cleveland acknowledged in 1931 that "the continuing squeeze on revenues which did not reach the inspectorate probably amounted to fifty percent."[63]

For the period 1914–1937, Lianghuai and Changlu together generated almost half of the national salt revenue. The former averaged 34 percent, followed by the latter which, except for the turbulent years of 1926–1931, averaged 12 percent. Despite all the modern reforms, the traditional revenue farmers and their various guises in these two divisions thus continued to play a major role in generating revenue for the state. To be sure, free marketing of salt enjoyed success to varying degrees in Sichuan, Yunnan, and Guangdong provinces, but attempts elsewhere switched back and forth between hereditary or *pro tem* revenue farming as well as government operation.[64] One need not criticize Dane and his colleagues for their attempt in imposing new versions of modernity on the gabelle system. Despite their determination, the institutions they constructed, as with the old, faced the fundamental dilemmas of cost versus revenue as well as stability versus maximization, and the state's ability (or inability) to control either. As anachronistic and nonrational revenue farming might be to its critics, the system remained, in the words of a not unsympathetic Gale, "truly one of the most ingenious fiscal systems in the world."[65]

Placing where credit is due, then, Dane and his foreign colleagues' contribution to modernizing the morass of salt administration in China should not be exaggerated.[66] To the extent that local and provincial salt revenues came under the jurisdiction of the central government with the assistance of foreign inspectors, the magical growth in audited salt revenue during the republican period may be traced in part to successive rate increases and the ability of revenue farmers in the major divisions to advance the gabelle.

While vulnerable to power politics and the demands of the groups consortium, the efficacy of the revenue inspectorate as an insulated institution also declined as the country descended into political chaos. Provincial and local withholding began to appear in 1916 with Guangdong, Yunnan, and Sichuan under Nationalist control, followed by various provincial authorities as the central government disintegrated after the death of Yuan Shikai. Between 1914 and 1937, while the total amount of gabelle audited grew consistently through rate hikes and other reforms, tripling from 68 million yuan to 217 million, salt released under audit for domestic consumption hovered between 37 million *dan* (1914) and 48 million *dan* (1922). Gale, too, considered 1922 the high point of the inspectorate, although gabelle surplus (i.e., after expenses and contribution to the sinking fund), made available to the central government, began to decline after 1918. The inspectorate's functionality increasingly depended, in addition to generous pay and benefit packages, upon the threat of foreign gunboats, the integrity and stoicism of its foreign and Chinese staff to negotiate the amount of withholding by provincial authorities at gunpoint.[67] Ultimately, it was the people, as consumers of salt, who bore the growing burden.[68]

Successive republican governments and the many local warlords they spawned thus derived a substantial portion of their revenue from salt as levies, with or without the approval of the Revenue Inspectorate. Surplus gabelle released by foreign banks went toward government expenses and servicing of more domestic bond issues and foreign loans, giving the foreign powers control over Beijing. Free-marketing of salt or meaningful gabelle reform thus mattered little. Instead, revenue security in the form of huge advances from privileged merchants remained critical. Even when policy changes were contemplated, the state, including its foreign employees, must consider the stability of its revenue. For better and for worse, Jiuda and the refined salt industry navigated through this functioning but increasingly dysfunctional salt institution: From Beijing (and Nanjing after 1927) to provincial governments and warlords; between the foreign-dominated inspectorate and the Salt Administration; and the republican state's often conflicting goals of revenue stability, gabelle reform, industrial development, and people's livelihood.

2
The Networks

The turmoil that modern China underwent shaped Fan Xudong's quest to save the country through industrialization. A native of Xiangyin, Hunan, he was the second son of a poor rural school teacher. After his father's death, he lived in a charitable home for chaste widows with his mother before joining his elder brother Yuanlian (1876–1927) at the Hunan School of Current Affairs (*Hunan shiwu xuetang*) and followed Liang Qichao into exile after the failed Hundred Days Reform in 1898. With financial support from Liang, Xudong attended Okayama Number Six High School (*Dairoku kōtō gakkō*) and Kyoto Imperial University. After the 1911 Revolution, he hurried back to China to work as an assayer at the Bureau of Currency (*Bizhiju*), Ministry of Finance, where he soon became frustrated with the bureaucracy. He became involved in the salt industry when Zhou Xuexi, in preparation for reforms under the terms of the Reorganization Loan, dispatched him to Europe to study gabelle administration and salt refining in 1913–14.[1] Hoping to practice what he had learned, Fan dropped his plan to pursue further studies in Europe and hurried home hoping to begin construction of the government-owned salt factory. The order, however, never came. With the prospect of reform dimming, he turned to private enterprise as instruments of change. Trained as an applied chemist, he was convinced that a strong China must have a chemical industry, beginning with refined salt.[2] Young and impatient, he deemed that a public opinion campaign to promote institutional change, as Jing Benbai had been waging, was too slow and its outcome uncertain. Defying Xiong Xiling's advice that the company could not possibly survive a battle with the powerful revenue farmers, Fan resigned from his well-paid government position to devote himself to the new enterprise.[3] Inspired by Fan's missionary-like zeal,

Jing, too, ignored Liang Shiyi and Zhang Jian's well-meant warning of a path fraught with obstacles.

This chapter reconstructs how Jiuda challenged the established salt institutions and business. To Fan and his colleagues, the system was at odds with notions of a strong state, free, and competitive market. As Schumpeterian entrepreneurs, they transformed salt, a commodity since ancient times, into an instrument of social, economic, and institutional change.[4] Beginning with company registration, they battled through the bureaucratic maze of Beijing while raising the needed capital. Their goal might be laudable, but the risks and uncertainties did not endear the company in the capital market, especially after Fan and his colleagues decided against guaranteed dividend (*guanli*) in Jiuda's public offering. They found the popular practice, designed to attract investment in limited liability corporations by turning shareholders into bondholders regardless of the company's profitability, unsound.[5]

Jiuda overcame, to the extent possible, these obstacles by managing its network portfolio of various social, institutional, and cultural resources that "somehow" forged embedded relationships.[6] By identifying these attributes, whether it be the ten mutualities (*shitong*)— lineage, surname, native place, classmate, schoolmate, work, business, shared goal, common foe, and life experiences—or the five affinities (*wuyuan*)—kinship, spatial, religious, associational, and material—we would be able to reconstruct the hierarchical and heterarchical structure of the resulting network (*guanxiwang*), how it operated through connectivity, reciprocity, multiplexity, and transitivity, as well as its limitations.[7]

To some business historians and economists, this is, of course, yet another example of what went wrong in modern China's economy. Bypassing the market through such particularistic ties, if not corruption, it is a recipe for inefficiencies and limited growth, if not failure.[8] On the other hand, recent research on emerging firms elsewhere has identified similar strategies to obtain resources critical for survival and expansion.[9] By removing some of the stigma, and bias, against networks, Jiuda's experience suggests that, at least in its inception, the strategy was a useful and rational response to solving the obstacles in its path. However, as much as Fan and his colleagues managed market, hierarchy, and network skillfully, the strategy also involved "dark sides" of social capital: uncertainties, maintenance costs, and vulnerability to transitive liability—a foe's friend could be fair game.[10]

Reconstructing the Network

An analysis of Jiuda's earliest list of eighteen shareholders subscribing a total of 415 shares at 100 yuan each and another list of forty-two with 443 shares dated December 1915, supplemented by a 1941 list of 663 shareholders (with their addresses and certificate numbers) totaling 13,745 shares offers a clue to how the company was embedded in a national network of officials, politicians, and investors. Many of these early shareholders (certificate numbers 1–200 on the 1941 list) could be identified with shared similar life experiences, cemented by overlapping attributes such as kinship, native place, school, and associational ties.[11] The number of shareholders (identified and unidentified) and the number of shares they represented are summarized in Table 2.1.

Although Fan Xudong and Jing Benbai committed themselves to raising 25,000 and 5,000 yuan respectively of the 50,000 yuan of capital in the company's first preparatory meeting, neither was affluent (see Appendix 1 for the company's capital, reserves, and profits over the years).[12] Jing's name did not appear on the first subscription list. Fan subscribed forty shares at 100 yuan each, his relative Li Binshi fifty shares, and Zhou Zuomin (1884–1955), then a bureau chief at the Ministry of Finance and Fan's close friend since their student days in Kyoto, five shares, making a total of 95 (or less than 23%) of the 415 shares in question.[13] Other pledges came from an assortment of Jing Benbai's relatives and associates: his relative by marriage Wang Daxie (style: Botang, who succeeded Fan Yuanlian as Minister of Education, twenty shares).[14] There were also fellow gabelle reformers such as Chen Jingmin (a native of Guizhou and vice-speaker of the House of Representatives, 1912–1916, forty shares.)[15]

The largest bloc of pledges came from Liang Qichao, his family, students, and known associates who pledged a total of 255 shares (or over 61%), as shown in Table 2.2.[16]

Table 2.1 Identified[a] and unidentified shareholders lists, 1915

	Identified	# Shares	Unidentified	# Shares
Ca. Feb., 1915	18	415	0	0
Dec., 1915	34	421	8	22

[a]Identified is defined as those ascertained with at least one of the network attributes such as kinship.

Table 2.2 Liang Qichao and his network in Jiuda, early December 1915

	Pledged	Paid-up
Liang Qichao	60	20
Liang Zhongrong (brother)	10	5
Zhou Xizhe (Liang's son-in-law)	5	1
Huang Mengxi (Liang's associate)	60	30
Xu Fosu (Liang's associate)	40	30
Fan Bingjun (Liang's associate)	20	3
Tang Rui (Liang's associate)	15	1
Tan Dianyu (Liang's associate)	10	
Deng Xiaoke (Liang's student)	20	5
Cai E (Liang's student)	10	
He Qingyi, with Mai Gongli (Liang's student)	5	

As an advocate of nationalizing the salt trade who preferred partnerships to limited liability joint-stock companies, Liang nevertheless encouraged Jiuda's establishment and the participation of merchants in politics.[17] The Fan brothers were family too: Yuanlian counted among his leading pupils and Xudong was entrusted with risky tasks such as arranging for Cai E's escape to Yunnan and smuggling Liang's manifesto against Yuan Shikai's monarchical ambitions to Shanghai for publication.[18] In support of Jiuda, Liang pledged sixty shares while his brother and son-in-law another fifteen shares. As Xudong recalled fondly, every time they met, Liang asked about the company's progress and tallied up the latest total share subscription.[19] Liang's other students and close associates together subscribed a total of 180 shares. A network of overlapping family, close friends, and associates that straddled provinces and politics came together to support the risky enterprise.

Although many of the original subscribers did not meet their pledges in full, the list of paid-up shareholders reaffirmed the pattern. In addition to Li Binshi (20 shares), Fan Xudong (20) was joined by his brother-in-law Zhang Qinji (10), making it a total of fifty out of 443, or approximately 12 percent of the subscribed shares.[20] Jing took ten shares, his brother, and his brother-in-law Wang Zhongyu, twenty-five shares. Daxie, a cousin of Kangnian and uncle of Zhongyu, appeared on the 1941 list as the holder of certificate number 67 with ten shares. Liang Qichao subscribed twenty shares, his brother Zhongrong five shares, his son Siyong two shares, his son-in-law Zhou Xizhe one share, and together with Xu Fosu (30), Huang Mengxi (30), Fan Bingjun (3), Tang Rui (1), and Deng Xiaoke (5), a total of ninety-five shares (22 percent). Together, Fan, Jing, and the Liang's strong network of family and close friends held over 40 percent of the company.

Centered around Fan, Jing, and Liang, an extended network based on weaker ties—friends and associates of more diverse backgrounds reinforced the core of the company's shareholders.[21] Jing's associates and fellow Zhejiang natives Ye Kuichu held three shares and Jiang Mengpin ten shares, while Fang Jilin, a bureau secretary at the Ministry of Justice and a Jing classmate, another twenty shares.[22] Zuo Shuzhen, an expert on salt matters and member of the Salt Administration Discussion Society led by Jing, subscribed five shares.[23]

Social and economic interests also transcended political persuasions in Jiuda's shareholder network. On the December 1915 list, Yang Du (1875–1931) appeared as the largest individual shareholder of Jiuda with fifty shares. Although he was one of the masterminds behind Yuan Shikai's monarchical bid, he had studied in Japan and, as an activist in the late Qing constitutional movement, worked with Xiong Xiling, Fan Yuanlian, Liang Qichao, Liu Kuiyi, and Zhang Qinji. Other officials included Qian Jinsun (Ministry of Finance, Salt Administration section chief; 3 shares); Yao Yongbai (Ministry of Finance, currency section chief; 1) Luo Zhenfang (vice-minister of Agriculture and Forestry, Acting Minister, 1913; 5); Liu Kuiyi (a leading member of the Tongmenghui in Japan and Minister of Commerce and Industries, 1912–13; 10), and Xiang Ruikun (appointed special adviser to the Bank of China in 1912 by Wu Dingchang; vice-minister of Commerce and Industries, 1912–13; 30). Parliamentarians Huang (30) and Chen Jingmin (20) were joined by others such as Deng Rong (member, senate, 10); Wang Jiaxiang (president of the senate; 2), and Li Xiaoxi (style: Jifang; house member, 2).[24] Together, they held over one-third of the company's shares. From Jiuda's inception, its shareholder network thus centered around the families, friends, and associates of Fan Xudong, Jing Benbai, and Liang Qichao and beyond, although commonalities that forged the embedded ties to Jiuda's network did not mean homogeneity of its members.

Network Structure

Indeed, the strength and resources of this network—individuals linked to each other by a variety of attributes overlaid by their public position outside the company—was also differentiated geographically as well as by the size of their holding in the company. Of the eighty-four early shareholders (share certificate number 1 to 201) who held a total of 1,524 shares, 86 percent came from Tianjin (484 shares), Beijing (464), and Shanghai (367), with the remaining from scattered locations around the country, including Wuxi, Hangzhou, Suzhou, Nanjing, Wuhan, and Canton. Thirty-four shareholders with 325 shares remain unidentified (an average of 9.5 shares). On the other hand, the fifty shareholders who could be identified by at least one network attribute (family, friend, native place, or one such) held an average of 23.9 shares each.

Jiuda's bank network was similarly embedded and differentiated. For much of its history, Jincheng Bank (founded in 1917 with head office in Tianjin until 1936 when it moved to Shanghai) served as the company's house bank in a close relationship, cemented by interlocking directorates. Through Jincheng's various branch offices and in his own name, Zhou Zuomin, the bank's general manager, eventually held 369 shares (2.7%) of the company.[25] He and Ye Gengxu (certificate #5, 30 shares), the bank's adviser, eventually served on the company's board. In turn, Jing Benbai and Fan Xudong, themselves shareholders of Jincheng, served successively on the bank's supervisory committee.[26]

But the ties between Jiuda and Jincheng went beyond mere cross shareholding. Zhou, Ye, and Fan Xudong's brotherhood began during their student days together in Japan.[27] At a time when Jiuda's capitalization stood at 50,000 yuan, Zhou had already extended the company a credit line twice as much, and repeatedly approved overdrafts and loan renewals over the objections of his more cautious managers. Unlike his often intractable treatment of other clients, Zhou boasted that no issue or loan, however big, between him and Fan would take over thirty minutes (twenty according to Fan and ten according to Yu Xiaoqiu, Fan's chief aide) to settle. Even though Jiuda was still paying market interest, Fan need not waste his time and energy, surely a transaction cost, in protracted negotiations with "ulcer-ridden bankers" who were hungry for business but yet would not bite.[28] From Zhou's perspective, loans at market rate was just regular business, but the bank also reaped a public relations windfall for supporting a company with a reputation of spearheading the national goods (*guohuo*) movement.[29]

Through Jing Benbai's Zhejiang connection, Jiuda's early investors also included Ye Kuichu, president of the National Commercial Bank. A well-known banker who served briefly as acting supervisor of the Da Qing Government Bank in 1911, Ye sought frequent advice from a close family friend, Wang Kangnian (editor of *Shiwubao* in the Qing dynasty with Liang Qichao, cousin of Wang Daxie, and Jing Benbai's maternal uncle).[30] Together with Jing, they were active in the Zhejiang railroad recovery movement led by Zhang Jian. Jing himself designed the bank's accounting system.[31] In support of Jiuda, the bank offered a credit line of 50,000 yuan with a personal guarantee by Chen Jingmin, another Jiuda shareholder.[32] For these reasons, Jing considered the bank reliable and urged Fan Xudong against overdependence on Jincheng and Zhou Zuomin whose many interests and political ambitions might imperil the company.[33]

Wu Dingchang's Yien Yieh Commercial Bank (Yanye yinhang, founded in 1915 in Tianjin) also held shares in both Jiuda and the Jincheng Bank (in which he was a member of the supervisory committee and board). He shared not only a similar background with Fan and Zhou as students in Japan, but also a common foe in Zhou Xuexi.[34] When Wu assumed control of a national newspaper, *l'Impartial* (*Dagongbao*), both Fan and Zhou became shareholders and contributed regularly to its pages, promoting reforms of the gabelle system and advocating a national industrial policy.[35] In addition, Wu and Fan served together in various organizations such as the China Industrial Promotion Association (*Shengchan xiehui*) and as trustees of the Nankai Institute of Economics.[36] However, as a specialized bank with revenue farmers as major investors, Wu's involvement was limited when it came to doing business with Jiuda.

In part due to its status as a branch of a quasi-state bank accountable to its head office, the position of the Tianjin office of the Bank of China in Jiuda's banking network was professional, if not peripheral. The bank owned no shares of the company and did limited business by providing a secondary guarantee of Jiuda's promissory notes for gabelle payments. When Fan Xudong approached the bank for more business, Bian Baimei, manager of the Tianjin office, limited its involvement, deferring to Jincheng's association with the company.[37] Bian's aloofness (or professionalism?), of course, might also be explained by his competing ties and allegiances. He had drawn on Zhou Xuexi and Sun Duosen, both old family friends, to launch his banking career at the Bank of China.[38] When Beijing politics shifted, Bian along with Sun moved to Tianjin and, as a board member, helped form the Zhongfu yinhang (Chung Fu Union Bank, founded 1916 and headquartered in Tianjin until 1930). At the core of the Sun family corporate empire, the bank financed Tunway Enterprises, one of modern China's earliest holding companies, with extensive interests in revenue farming, flour mills, and Tongyi Salt Refinery (which became Jiuda's main competitor/partner), as well as 39 shares (0.28%) of Jiuda by 1941. Despite these competing ties, Bian maintained a close personal relationship with Fan, frequently sharing information, advice, and access, in addition to arranging his chemist sons to work for Fan.[39] In practice, then, Jiuda's bank network was also internally differentiated by the varying strengths of Fan's relationship with the banks and the bankers, as well as the bankers' own rival claims, creating a hierarchy within Jiuda's bank network. Other banks, without access to Fan, would have to find pathways to secure Jiuda's business.[40]

Network Resources, Operation, and Management

The heterogeneity of Jiuda's network highlights its multiplexity and transitivity.[41] A node in his own right as a leading intellectual and national political figure, Liang worked with (and against) Yuan Shikai in various capacities, including brief stints as Minister of Justice (1913–1914), director of the Bureau of Currency Reform (*Bizhiju*, 1914), and the Council of State (*Canzhengyuan*, 1914). In the last position, his colleagues included fellow company shareholders Li Yuanhong, Wang Daxie, Wang Jiaxiang, Chen Jingmin, Deng Rong, Cai E, as well as old associates such as Yang Du and Xiong Xiling.[42]

Liang also reinforced the network by serving as a structural hole providing other members of the network access to the highest level in the Beijing government through Liang Shiyi, a fellow Cantonese then serving as secretary of the president's office.[43] Liang Qichao's recommendation of Xu Fosu through Shiyi was accepted by Yuan Shikai, and Xu became a councilor (1914–1915) at the State Department. Liu Kuiyi, another shareholder, similarly approached Liang Shiyi indirectly through Qichao.[44] Shiyi himself became a shareholder with a 6,000 yuan stake in the company by 1918.[45]

Native place ties and shared life experiences also reinforced the network's cohesion. Fan Xudong, Xiong Xiling, Yang Du, Xu Foshu, Fan Bingjun, Li Binshi, Li Xiaoxi, and Liu Kuiyi all hailed from Hunan. From Zhejiang came Jing, Fang Jilin, Jiang Mengping, Luo Zhenfang (brother of Zhenyu), Mo Shumo, Tang Rui, Tang Peisong, Wang Jiaxiang, Wang Daxie, Ye Kuichu, and Zhang Hu.[46]

Many of the shareholders shared common ties: Jing Benbai and Fang Jilin, for instance, were also classmates (*tongnian*) in the civil service examination; Cai E, Li Binshi, and Fan Yuanlian were schoolmates at Hunan shiwu xuetang in China; and Cai E, Chen Jingmin, the Fan brothers, Huang Mengxi, Li Xiaoxi, Liu Kuiyi, Wang Jiaxiang, Xiang Ruikun, Yang Du, Yao Yongbai, Ye Xugeng, Zhang Qinji, and Zhou Zuomin, for example, all studied in Japan.

The expanding network was also reinforced by common political foes and overlapping affiliations. In addition to his opposition against gabelle reform, Zhou Xuexi was instrumental in ousting Jing Benbai, Zhou Zuomin, Zhang Hu, and Wu Dingchang from their offices. Wang Jiaxiang (vice-president, 1912 and president of the senate, 1916), Chen Jingmin, Li Xiaoxi (a member of the Tongmenghui), and Deng Rong from Sichuan were shareholder-parliamentarians spanning various political parties. Both Wang and Chen were staff members (*ganshi*), along with Fan Yuanlian and Zhang Hu, of the Republican Party (Gonghedang) headed by Li Yuanhong, Zhang Jian, and secretly funded by Yuan Shikai. Jing Benbai was active in the National Alliance (Guomin xiehui), which merged with the Democratic Party (Minzudang) led behind the scenes by Liang Qichao. Huang Mengxi (style: Dajian) was a member of the Unification Party (Tongyidang) under Xiong Xiling, Zhang Jian, and Zhang Hu. The three parties eventually merged in 1913 as the Progressive Party (*Jinbudang*) led by Liang Qichao to form a parliamentary majority in support of Yuan Shikai.[47] On the other side of the political aisle, Liu Kuiyi, a longtime member of the Tongmenghui and the Nationalist Party served briefly as minister of commerce and industry, and headed his own party, the Comrade Society (Xiangyouhui).[48] Han Yongqing, holder of certificate number 64 (100 shares) on the 1941 shareholder register, made his fortune as a compradore for the International Export Company and served as Sun Yat-sen's adviser.[49] The company's shareholders thus straddled the three main political camps in the early republic: Yuan Shikai, Liang Qichao's Progressive Party, and the Nationalists.

These officials, politicians, and bankers brought more than mere monetary capital to the company. Instead of being passive equity holders of Jiuda, they provided a variety of resources while lending their prestige and influence in promoting and protecting the company.[50] On the lists were cabinet ministers: Liang Qichao (justice, 1914; finance, 1917), Liu Kuiyi, Xiang Ruikun (commerce and industries, 1912–13), Wang Daxie (education, 1913–14; transportation, 1916; and foreign affairs, 1917), and Luo Zengfang, (agriculture and forestry, acting,

1913). Zhuang Yunkuan (1866–1932, holder of certificate 67 with ten shares) served as the chief censor (1914–1916) (*Du suzhengshi*) and headed the Audit Yuan (*Shenjiyuan*) from 1917 to 1927. In addition to Liang Qichao's brief service as director-general of the Salt Administration in 1917, at least two divisional commissioners were also involved: Li Binshi in Changlu (1913–1914), Northeastern provinces (1915) and Liangzhe (1915); Luo Zhenfang in the Northeastern provinces (1913–1915) and Deng Xiaoke who reformed Sichuan's salt administration system, as well as members of Jing Benbai's Salt Administration Discussion Society: Chen Jingmin, Fang Jilin and Zuo Xiqin. Jing's associates from the Zhejiang railroad rights recovery movement and the National Commercial Bank—Tang Shouqian, Ye Kuichu and Jiang Mengping—also provided support, as well as Liang Qichao's close associate Tang Rui at the Bank of China (acting vice-chairman and chairman, 1913–1914).[51] Without the support of these relatives, friends, and associates, as one employee recalled, the company would not have gotten off the ground.[52]

The Politics of Company Registration

With the resources of this network, Jiuda battled the new synarchic salt bureaucracy for a toe hold in the salt market. The technology for refined salt was not demanding, but the company encountered obstacles that made its registration far from routine.[53]

In crafting their petition to register Jiuda, Jing and his associates carefully avoided the issues of revenue farming and divisional boundaries. Instead, they appealed to patriotism, of restoring national health, both fiscal and hygienic. The flood of foreign salt, not merely as contraband but also tax-free, must be met at the "commercial ports" (*tongshang kou'an*).[54] Jiuda's new refined salt would also provide an alternative to and set an example for traditional salt producers. For far too long, the health of China's citizenry had been jeopardized by "eating dirt"—the crude salt produced by solar evaporation contained much sand and other impurities. The company proposed a refinery at Tanggu using modern methods: brine pumped through filters, treated with sodium carbonate, and then boiled to produce hygienic salt, powdered or bricked by machine, with revenue stamp affixed on the paper packaging. Pleading that a full gabelle would doom the young company in its competition against tax-free foreign salt, the applicants argued that the government should promote this modern industry by waiving half the tax for three years. Just as other modern domestic industries were protected in their supply of raw material and labor, no other refinery should be permitted within one hundred *li* of Tanggu.⁵⁵

Jiuda's appeal to patriotism and consumer welfare, however persuasive, failed to impress the bureaucrats. As outlined in the previous chapter, the economic, political, and institutional reality of the early republic limited the possibility and scale for reforms. While Dane believed that competition would motivate producers to improve the quality of their salt and cutting prices, his inspection tour had convinced him that solar evaporation was the appropriate technology for China. Endorsing Beijing's request to divert to other uses half of the 20 million yuan earmarked for salt-related reforms, he noted that investing 3 million yuan of the Reorganization Loan in a salt factory in North China was a waste.⁵⁶

Jiuda's petition thus ran into skepticism and bureaucratic hurdles. The Revenue Inspectorate ruled out any tax waiver. The gabelle had been pledged to repay the Reorganization Loan, making an exception for the company impossible. Guarding its revenue flow jealously, the inspectorate also demanded clarifications: the authority responsible for issuing the revenue stamp and hence the revenue. Jiuda's planned source of raw material—whether in the form of brine or crude salt— also created headaches.⁵⁷ How might brine be taxed and, if so, at what rate? Should taxation of crude salt as an alimental commodity be distinguished from its use as an industrial raw material?

Fortunately for Jiuda, Jing Benbai was well-placed to negotiate the bureaucratic labyrinth. Although he did not enjoy Dane's veto power, Jing, while serving as domestic adviser to the Salt Administration, had drafted the Presidential Special Order on Salt Production, promulgated on March 4, 1914. Breaking the hold of hereditary saltern households and revenue farmers on salt production and wholesaling, clause 1(D) of the order specifically authorized the production of refined salt by any duly registered company.⁵⁸ Supplemental regulations issued by the Finance Ministry on September 9, 1914, further specified that only Chinese nationals could engage in the industry, and that company articles of incorporation must be submitted to the Salt Administration for approval.⁵⁹

With the legal groundwork prepared, Jing filed another petition in August 1914 addressing the questions raised by the Revenue Inspectorate. The finance ministry would issue the revenue stamp in lieu of transport permit (a costly procedure under the Qing) not merely to simplify the process but also to verify gabelle payment and distinguish the company's product from crude salt. Jing also argued that using brine as the source of raw material was more efficient and hygienic. Solar-evaporated crude salt purchased from the state depot must first be dissolved and filtered, which added to the time and cost of manufacturing.[60] Those disadvantages, however, did not concern officials, who focused only on ease of revenue assessment and collection. Without ascertaining the salt content (and weight) in the brine, proper gabelle assessment was impossible. The only possibility, then, was crude salt, which also posed another problem. Established regulations required that the gabelle must be paid in full before release from the state depot, which meant that Jiuda was also paying tax on dirt, sand, and other impurities. Jing then pleaded that the company be permitted to purchase its raw material directly from saltern households on a long-term contract to gain some measure of quality control over its salt supply and pay the gabelle on the weight of the finished product only when shipped. Last but not least, as one of the company's goals was to compete against tax-free foreign imported salt, paying the gabelle even at half the full rate would be an additional boon to state coffers while promoting domestic production and industrial development.[61]

The Salt Administration's tentative approval of Jiuda's registration dated September 22, 1914, acknowledged the company's innovative product and accordingly granted Jiuda a geographical monopoly in salt refining within a hundred *li* of Tanggu. After consultations with the Revenue Inspectorate, however, the request for partial gabelle waiver was denied again: the inspectorate reasoned that unlike the flour and textile industries, salt did not face foreign competition, the company's claim notwithstanding. However, as a concession and exception, the inspectorate agreed to accept Jiuda's gabelle payment in promissory notes instead of cash, with the length of the grace period to be decided later.[62] The company did not win all its arguments with the inspectorate, but it was registered officially. Fan bought land at Tanggu, prepared the site, designed the plant, conducted experiments, traveled to Japan to purchase boilers, and then on to Shanghai for the evaporation pans. By October 1915, the plant was ready for production.

Battling Bureaucrats

Jiuda's battle, however, had only just begun. The return of Zhou Xuexi in March 1915 as Finance Minister exposed the transitive liability of the company's network. Once back in office, Zhou orchestrated

Zhang Hu's removal on a charge of corruption, which also forced Jing's resignation as adviser to the Salt Administration for having shared Zhang's graft by accepting a subsidy to his journal *Yanzheng zazhi*.[63] Henceforth, Jing continued his campaign as a private citizen, albeit a well-connected one, over issues that affected the company's viability: the length of grace period and method of gabelle payment, franchise regulations, scientific determination of wastage rate, and whether the gabelle should be adjusted accordingly.

The war of words continued over the terms of gabelle payment. Jiuda, the Revenue Inspectorate insisted, must pay the gabelle in full before crude salt, dirt and all, could be released from the state depot. Again, Jing argued that an exception should be made. The company must process the salt before refining, unlike salt merchants who could begin selling without further processing. To make good its losses in time and material, there should be a six- to eight-month grace period for the promissory notes. On August 31, 1915, the inspectorate finally conceded the point and agreed to a four-month delay, provided that such notes were secured by domestic government bonds up to five-sixths of their market value.

The Revenue Inspectorate's proposal seemed reasonable in protecting the revenue flow to service foreign loans, but nonetheless an impractical solution from Jing's perspective. Unlike salt merchants who obtained release in one consignment what they needed for the entire season, Jiuda's continuous refining process required only delivery on time, not bulk. Furthermore, even if the officials involved could process the paperwork, the company deemed it too cumbersome to make daily visits to the Changlu Inspectorate with the promissory notes, fix the redemption date, and negotiate the value of domestic bonds in the absence of a national bond market. Instead, Jing countered that the company be permitted to use forward-dated promissory notes payable in six months.

In his recommendation, C. H. Lauru, the inspectorate's French financial section chief, conceded "good grounds for simplifying the mode of payment." He recommended that Jiuda's promissory notes should be guaranteed by the Bank of China. The risk of Jiuda's default would thus be transferred to the bank, and determination of the necessary security (as well as handling and interest charges) a matter between the company and the bank. For accounting purposes, the inspectorate could treat such papers as cash in hand, facilitating "greatly...the keeping of accounts and returns of salt issued." To ensure a timely flow of revenue, however, he remained unmoved on the issue of grace period. Ernst von Strauch, the German associate chief inspector, and Gong Xinzhan, Chinese chief inspector, approved Lauru's recommendation.[64]

Defining the Market

Nevertheless, the fate of Jiuda still hung in balance with Zhou Xuexi in office. Before it could begin shipping, the company submitted in July 1915 to the Salt Administration its franchise regulations. These were rejected for two sensitive issues: the definition of "commercial ports" and how revenue farmers who held the privilege to sell salt in those ports—however defined—might be affected.

The astute Zhou, of course, understood the threat Jiuda posed to the revenue farming system. The young company needed markets, the larger the better, not merely against foreign import, but also to breach the monopoly distribution system imposed by revenue farming. Without recusing himself because of his family's substantial revenue farming interests in the Lianghuai division, Zhou ruled that the company's market be limited to only those foreign concessions found in the "treaty ports" and then only through the exclusive agency of existing revenue farmers, thus confining Jiuda to the thirteen treaty ports with foreign concessions and an additional twenty-five resorts and territories where foreigners could reside by treaty. Turning economic nationalism against Jiuda, Zhou reminded the petitioners of their avowed goal to compete against tax-free foreign imports. To live up to its words, the company should not quibble with his stringent definition. In addition, duty stamps must be affixed to the packaging at the rate of one stamp per catty, with the company responsible for the printing cost.[65]

True to his reputation as a wizard of words, Jing insisted that the term "commercial ports" should be broadly defined to include not merely those with foreign concessions and residence rights, but also any ports declared open to foreign trade by the Qing and Republican governments (*zikai shangbu*).[66] Jiuda would thus enjoy access to all seventy-seven treaty ports with or without foreign concessions and an additional twenty-eight declared ports. His masterful choice of words put Zhou and the ministry in a quandary. Equating "commercial ports" as treaty ports implied compromising Chinese sovereignty over all the ports declared open for foreign trade. On the other hand, restricting trade only to those ports with concessions or right of residency would also incur the wrath of various foreign powers.[67]

Jing filed conciliatory appeals offering a way out for Zhou and his secretaries.[68] He argued that Jiuda could not possibly compete against revenue farmers with the heavy gabelle carried by refined salt. The company's minuscule output capacity of 30,000 *dan* per year would not interest revenue farmers handling over 40 million *dan* in annual sales nationally. Furthermore, with a commercial port often shared among several revenue farmers, dividing further the limited supply of refined salt with its slim profit margin, if any, gave them no incentive to promote the new product against tax-free imported salt. Given the uncertain prospect, asking them to sell a fixed quota was unrealistic, if not unfair. He proposed that existing revenue farmers be offered a first right of refusal as sales agent (*daixiao*), as opposed to a franchisee (*baoxiao*) who must fulfill a fixed quota. The company would offer existing revenue farmers the opportunity, valid for two months, after which it would be free to contract with other parties or open its own outlet. Jiuda also promised existing revenue farmers a first right of refusal again if and when the company's sales of refined salt exceeded 10 percent of the market. As for the need of revenue stamps, he reminded the officials that at the rate of one revenue stamp per catty, the company needed to pay for 6 million stamps. Caught in the web of definitions and unwilling to let Jiuda loose, Zhou and his secretaries at the Salt Administration decided on the best course of action for bureaucrats—doing nothing.

How the impasse was broken remains opaque. Fan Xudong initially approached Yan Xiu, tutor to Yuan Shikai's sons and Yuanlian's patron, for assistance. However, Yan, whose household also owned and operated for generations the Changlu monopoly of Sanhe as a revenue farmer, declined to intervene as he disagreed with Yuan's monarchical ambitions.[69] Fan then appealed to Li Shiwei (style: Bozhi, chairman of the Bank of China) to intercede on behalf of the company. The effort soon bore fruit when he was summoned to an interview with a sympathetic Li Sihao, chief of the Salt Production Bureau, who promised to brief Zhou again.[70] Jing Benbai credited the company's success to Yuan Shikai's increasing dissatisfaction with Zhou Xuexi.[71] The final push, however, allegedly came from Yang Du who secured Yuan's approval over dinner with a sample bottle of the company's refined salt.[72] Following Zhou's resignation and Zhang Hu's brief return as vice-minister of finance, succeeded by Li Sihao on May 20, 1916, the company's franchise regulations received provisional approval five days later. Jiuda was still responsible for the printing cost of the revenue stamps, and its annual production was officially capped at 30,000 *dan* of salt to be marketed at various commercial ports, however defined. Existing revenue farmers, if any, had two months to decide whether or not to become the company's agents.

Wastage

One final hurdle remained. Conceding defeat on a partial gabelle waiver, Jing tirelessly pursued the issue of wastage.[73] By law, the company must purchase its raw material from the Changlu state depot and pay a gabelle of 2.75 yuan per *dan* of crude salt. Conducting its own chemical analysis, summarized in Table 2.3, the company found that the "salt" it bought contained over 30 percent of moisture, dirt, and other insolubles by weight.[74]

The implication for Jiuda's production cost with sodium chloride set at 98 percent was not insignificant as seen from Table 2.4.

Table 2.3 Chemical analysis of Changlu crude salt

NaCl	67.46	Percent
Gypsum and limestone	5.75	—
Moisture	14.62	—
Insolubles	12.17	—
Total		100

Table 2.4 Cost analysis of refined salt (NaCl at 98%)

Effective gabelle	3.9[a]
Salt cost	0.28
Wood crate	0.2
Paper wrapping	0.3
Packaging	0.1
Production	0.8
Transportation to Tianjin	0.2
Administration/advertising	0.8
Total cost	**6.58**

[a] Proportionately adjusted for wastage at the nominal rate of 2.75 yuan/*dan*.

With the effective gabelle rate for Jiuda's refined salt almost 42 percent above the nominal rate, Jing Benbai argued that a distinction should be made between alimental crude salt as a finished product sold by revenue farmers from salt as an industrial raw material for further refining. Without a wastage allowance, a *dan* of refined salt cost 6.58 yuan to produce. A Tianjin market price of 5.6 yuan would mean a loss of 0.98 yuan.[75] Surely, determining the percentage of wastage and purity of sodium chloride was a scientific matter. The company thus petitioned for tests and adjustment in calculating the gabelle accordingly, if only to fulfill its patriotic mission.[76] Official chemists, both foreign and Chinese, confirmed the 30 percent "loss" in the refining process.[77]

However, Sir Richard Dane remained skeptical and stood firm as a matter of principle and policy consistency. The company was proposing a scientific but unprecedented method to calculate the gabelle—by weight of sodium chloride of a certain degree of purity. If revenue farmers and other dealers were to follow suit with similar claims to calculate the gabelle only according to the shipped weight of salt of a certain degree of purity, the impact on revenue could be substantial.[78] Simply put, it was an "established fact" that one *dan* of Changlu crude salt, dirt and all, carried 2.75 yuan in duty "at least receivable."[79] After repeated petitions, denials, and appeals stacked over a foot in height, the Revenue Inspectorate, through the good offices of Liang Qichao, Zhang Hu, and Li Sihao at the Salt Administration, finally conceded a 30 percent discount. Eventually, Jiuda came to control through vertical integration the quality of raw salt from its own salt pans (purchased as Fan Xudong's personal property), supplemented by saltern households contracted to produce for the company exclusively.[80] As modern a bureaucracy as the Revenue Inspectorate, scientific analysis of wastage, impurities, and public health, much less development of a modern industry, mattered little when pitted against its primary charge of revenue security.

From the perspective of neoclassical economics, business historians and economists have long been critical of networks: reliance on such "irrational" particularistic ties, if not corruption, in bypassing market competition, at-arms length relationships, and meritocratic control led to failure. Even though the business network is very much part of European and American business (both past and present), Asian business networks are still stigmatized (especially after the bursting of the Asian bubble) and considered an "Oriental" enigma.

However, Jiuda's experience suggests that the business network may be a useful and rational response to the challenges of doing business during the early republican period. In managing the many obstacles in its path, Jiuda drew on the resources—capital, information, and influence—of its shareholder network. A portfolio of strong ties centered on family, kin, and friend, reinforced by native place, school, work, shared life experience, and political allegiance as well as common foes, created a differentiated and hierarchical structure of relationships, which facilitated the company's capital drive and registration. Such substantive considerations are not mutually exclusive of profit maximization (or minimizing loss); rather, it was joint optimization. The art of management (and entrepreneurship) involved not merely matching output with demand, but also managing market, network, and hierarchy as well as the costs of network maintenance, limitations, dependency, redundancy, and transitive liabilities. In short, network enables as well as constrains.

3

Breaking into the Market

On May 13, 1923, Jiuda received from Li Yuanhong, one of its shareholders, a presidential citation for contributing to the people's sustenance (*minshi*) and over 2 million yuan in gabelle annually to the country's coffers.[1] This chapter analyses the reasons for Jiuda's success and how it continued to wrestle with the institutions of the republican government, law, and the market. At times Kafkaesque, the saga of its progress unfolded with ironic twists and turns. Disagreements between them notwithstanding, Jiuda was saved, at least in part, by the synarchic Revenue Inspectorate. The descent of the country into political chaos, too, created both opportunities and crises for the company. In addition to a skillful blending of market, network, and hierarchy, Fan and his colleagues championed free marketing of refined salt with petitions and lawsuits while exploiting the different interests and factionalism within the salt bureaucracy as well as the division between the central and provincial governments. Drawing on their networks, influence (*shili*) could be exerted on officials to do nothing or decide in favor of the company, reinforced by goodwill (*qingyi*) invoked through network transitivity or manufactured by rent creation.[2] When the strategy failed, the company resorted to bribery. To secure Jiuda's place in the heavily regulated and imperfect salt market, Fan's entrepreneurship skill includes not merely managing the company, but also mastery of the economics of regulation, rent extraction, and political extortion.

Economics of a National Distribution Network

With over 20,000 yuan already spent on land, physical plant, and machinery, Jiuda must find a way to solve two interlinked problems: raising more capital and financing a national marketing system.

The variety of gabelle rates applicable to different divisions shaped a partial solution. In early 1917, Jiuda tested the waters by applying for a gabelle refund for its shipment to divisions with a lower rate than Changlu. At stake was the feasibility of various markets for Jiuda. The company argued that it could not afford to ship its product, already with a higher manufacturing cost than crude salt, to markets with a gabelle rate lower than Changlu without at least a partial refund. Affirming the promotion of a hygienic product as one of its goals, the Revenue Inspectorate nevertheless denied the petition by a bureaucratic catch-22: just as the company was unwilling to ship at a loss, the government was also not prepared to discuss a "hypothetical question of refund without evidence of actual shipment, location, and gabelle paid." Despite the company's appeals, the Salt Administration and the inspectorate stood firm, claiming that the gabelle rates of various divisions cited by the company was "confusing and unclear."[3] An internal memo revealed another reason: the Revenue Inspectorate must not compromise its principal object of maintaining "the balance of demand and supply with no detriment to the interest of the government" by granting rebate or market access.[4] Fujian and Yunnan commissioners also declared that the market for refined salt was limited because their divisions already had an ample supply of crude salt.[5] Given these constraints, Fan Xudong and his colleagues could only concentrate on markets with a gabelle rate higher (e.g., the four Yangzi provinces) or equal to Changlu while solving the problem of capital shortage by blending hierarchy with network and market. Shipment of refined salt in bottles, paper packets, wood crates, and gunny bags to Jiuda's own outlet in Tianjin began on October 5, 1916. The location was strategic: on Liujili off East Avenue (*Dongmalu*), one of the main commercial thoroughfares in the old part of the Chinese (i.e., not in any foreign concession) city served by a tramway. In November 1917, another company-owned outlet opened at Xiaguan inside the area designated for foreign residence in Nanjing with a contract to supply 1,000 *dan* of refined salt monthly to the British-owned International Export Co., Ltd., an egg product and meat processor whose compradore, Han Yongqing, was also a major Jiuda shareholder.

Jiuda's shareholder network also shaped its marketing system elsewhere in the Yangzi provinces. With its limited capital and managerial manpower, the company contracted Hankou, its potentially most lucrative market to Xinda, through an unlimited liability partnership formed by the Jing brothers (Benbai and Jiye), Li Guanqing (certificate #50, 15 shares), Fang Jilin, and Ge Wangyun.[6] Official approval of the franchise came quickly from the provincial salt bureau where Jing Benbai found a sympathetic former colleague in charge.[7] Wholesaling operation was located at Qianhualou with a network of retailers throughout the city. Jiuda thus solved its problems while Xinda bore the costs and risks of developing the market.

Changsha was another targeted market, although Jing's efforts ran into opposition from both the provincial salt bureau and the established revenue farmers. Zhang Qinji, Fan's brother-in-law and fellow Hunan native (and Jiuda shareholder), intervened, lending his prestige and resources as a banker and an industrialist with substantial holdings in spinning mills, collieries, and tungsten and antimony mines scattered through the province. He formed Shengda, another unlimited liability partnership with a group of hereditary licensed holders as the company's franchisee. Zhang's extensive network of provincial politicians and capitalists thus facilitated Jiuda's expansion into that market with access to local capital and information.[8]

Where Fan and Jing could not find a suitable arrangement from within, they recruited someone deemed knowledgeable about local conditions from the market. The Nanchang franchise was offered to a partnership headed by Gong Shicai, chairperson of the Nanchang Chamber of Commerce and a banker, with its security deposit held as Jiuda's shares.[9] Even if Gong could not promote the company in public because he was embedded in a local network with revenue farmers among his relatives and friends, he promised to work behind the scenes through a proxy.[10]

Other forms of capital also formed the basis for a franchise. Anqing went to Wu Jinyin, brother of Wu Jinbiao, commander of the Northern Jiangxi garrison.[11] Jinyin, as a director of the company, and his brother provided critical access not merely to the local market but also to warlords and provincial officials.[12] For this special capital, Jiuda waived the security requirement, although it assumed control of the franchise's accounting. The Wu brothers and family invested in the company under a variety of names, eventually holding 423 shares in the 1941 register.

Although not without risks and franchise opportunism as predicted by agency theory, Fan and his colleagues thus built Jiuda's distribution system by blending hierarchy and network with market, and modern limited liability corporations with traditional unlimited liability partnerships.[13] In addition to company-owned outlets, Jiuda also used franchisees to lessen the burden on scarce managerial and capital resources in the fledgling company while benefiting from their local expertise.

Whatever the institutional arrangement, however, contracts specified the duties and responsibilities of the parties. The marketing area was clearly defined, and an interest-bearing security deposit required in most cases, or in lieu of that, surety in the form of real estate and other properties.[14] Payment for Jiuda shipments must be settled monthly. If the amount due Jiuda exceeded three quarters of the security deposit, the company reserved the right to suspend shipment. In the event of default, the company could seize the surety, as it did in the case of Zhang Qinji after his death, although he was Fan's relative.[15] In other words, market coexisted with network as Jiuda evolved two separate marketing systems, one controlled directly by the company hierarchically and another of partnership franchises.

Both systems served Jiuda effectively, its timing to break into the market opportune as declining Huainan output strained supply in the Yangzi provinces. Barely four months after opening in Hankou, the company petitioned to raise its production limit to 100,000 *dan*. As a result of the First World War and dwindling supplies, foreign troops stationed in China and exporters such as International Exports had turned to Jiuda for their needs. The Army and the Navy Ministries, too, made frequent inquiries about the supply of refined salt in brick form for military use. Industrial chemical manufacturing such as explosives and soda ash was also growing. None of these new demands competed with crude salt producers and merchants. With Liang Qichao as director-general of the Salt Administration (April 17, 1917 to November 30, 1917), Li Sihao as chief commissioner (May 20, 1916 to January 10, 1919), and Fan Yuanlian as acting Foreign Minister, concurrently Education Minister and head of the Interior Ministry (January 1, 1917–June 12, 1917), Jiuda's petition was quickly approved. By order #2164 in September 1917, the Salt Administration generously promised to raise the limit as exports grew.[16] By the end of the year, the company's production cap was raised to 150,000 *dan*, a fivefold increase from 1914, although still a tiny fraction of the 40 million *dan* being consumed in the country annually.

Revenue Farmers' Reaction

As Jiuda gained a toehold in the market, its war against Changlu and Lianghuai revenue farmers also began. An avalanche of petitions from the Changlu salt merchants' guild and Gongyi, holder of the Tianjin-Wuqing monopolies, opposed Jiuda's encroachment. Gongyi argued that neither the Salt Administration nor the company had the authority to violate its privilege as exclusive supplier to Tianjin. On August 12, 1916, the Changlu guild petitioned the Salt Administration to stop Jiuda, raising the thorny issue of defining "tongshang kou'an." The next day, Gongyi also petitioned the government for assistance in its negotiations with the company. Under the officially approved franchise regulations, Gongyi enjoyed the right of first refusal, but Jiuda's demand for a guarantee of 5,000 *dan* in annual sales presented too much risk for Gongyi to assume, not to mention a sizable security of 0.52 yuan/*dan* subjected to confiscation if sales fell short of the quota. Acknowledging Jiuda's goal of competing against foreign imports, the revenue farmer argued not unreasonably that market response to the new product (or foreign reaction for that matter) was beyond its control. As a private business matter, even if the state approved of Jiuda's franchisee regulations, the company should not unilaterally dictate those terms. Jiuda, on the other hand, insisted that Gongyi must accept those terms to ensure its due diligence in promoting the consumption of refined salt. Petitions and counterpetitions continued until the Salt Administration, then headed by Zhang Hu (a Jiuda shareholder) ruled on October 12, 1916, that Jiuda's franchise offer, duly served, had expired. The company was free to act.[17] Three days later, Jiuda's outlet opened for business and, by the end of the year, sold a total of 452.22 *dan* of refined salt. However, the company's annual sales in Tianjin did not reach 5,000 *dan* until five years later.

The Lianghuai merchants, too, waged war. A mutual friend allegedly approached Jing with tempting counteroffers: To buy Jiuda for 400,000 yuan, half of which Jing could keep for himself; or 30,000 yuan annually in subsidy to him, whether as the new head of the Lianghuai Guild or for his journal. Combining the carrot and the stick, the revenue farmers also threatened to invest 10 million yuan in a salt refinery (the Lequn Salt Refinery, see chapter 4) to crush the interloper. A defiant Jing rejected the tempting offers.[18] The deadline for the franchise offer expired without a contract.

The revenue farmers, however, had Lianghuai's commissioner on their side. Liu Wenkuai reminded his superiors in Beijing that he should not be accountable if the division failed to fulfill its gabelle target with this invasion. If competition against foreign imports was such a pressing issue, the revenue farmers would serve as Jiuda's agent with an annual quota of 6,000 *dan*, more than adequate for the small number of foreigners residing in the division. Liu pleaded that further negotiations be authorized by the Salt Administration. Beijing's rejection came swiftly. Jiuda had in July 1916 followed the franchise procedure as approved and a two months' notice duly served. Since the offer had expired, the government could not stop Jiuda in Hankou. Instead of procrastinating, the revenue farmers should embrace competition as an incentive to improve the quality of their salt.[19]

Lianghuai merchants did not surrender meekly, however. For five years the revenue farmers, Jiuda, and the Salt Administration continued their battle in the Administrative Court (*Pingzhengyuan*) over the definition of commercial ports, packaging methods, and production caps.

Definition of "tongshang kou'an"

This round of Jiuda's troubles began when it opened another outlet inside the west gate of Nanjing city with an advertising campaign to attract customers. What seemed a routine business operation provoked a furious response from Yihexiang, which, as the exclusive salt supplier to Nanjing, felt it had been cheated already of the International Exports contract. While the Salt Administration sanctioned Jiuda's outlet inside Xiaguan, any location beyond it was another matter.[20] Several issues were raised by the company's use and interpretation of the term "commercial ports" (*tongshang kou'an*): whether it was coterminous with "treaty ports" (*shangbu*) or "foreign concessions" (*zujie*); the precise boundaries of such designations (however defined); and the legality of consumers to purchase, transport, and consume refined salt beyond those boundaries. When queried by the Salt Administration, the Foreign Ministry offered a diplomatic answer in July 1917: the boundary of treaty ports had long been a matter of continuing dispute between China and the foreign powers. Whereas both the Manchu and the republican governments insisted on a strict definition, that is, areas designated for foreign residence and demarcated accordingly as concessions by treaty, the powers consistently sought a broader interpretation, one that included not merely the rights of residency and purchase of land but also to conduct trade anywhere within the port, that is, beyond the area designated for residence. Taking the query as directed against imported salt, the Foreign Ministry recommended a strict definition. Accordingly, the Salt Administration equated the term "tongshang kou'an" with "foreign concession."

Jiuda appealed the decision, reminding the Salt Administration that it was a duly registered Chinese company with a charter to serve "commercial ports," not treaty ports. Queried again, the Foreign Ministry acknowledged that its earlier recommendation was applicable only to foreign companies and provided another diplomatic solution: as a domestic issue, the Salt Administration could exercise its discretionary authority to adopt the broad definition. Accordingly, the Salt Administration issued order #939 on July 4, 1918, adopting a broad definition of "tongshang kou'an." The location of Jiuda's outlet inside the Nanjing walled city was not illegal.

This flip-flop only unleashed a flurry of more petitions from both sides while revenue farmers in Suzhou and Nanchang also filed pre-emptive suits.[21] In an attempt to stem the tide, an exasperated Salt Administration issued order #520 restricting Jiuda's outlet location henceforth to within the foreign concessions inside treaty ports, and where there was no concession, the area designated by the government for foreign residence. To resolve the issue in Nanjing, Yihexiang would be authorized to verify the company's shipments.[22] On the other hand, on December 9, 1919, the Salt Administration also dismissed a Lianghuai petition seeking to limit Jiuda's production and sales. As refined salt was aimed at competing against imported tax-free salt, the industry should be promoted. The orders again satisfied neither side.

Jiuda quickly challenged the Salt Administration's order #520. The order not merely conflated "tongshang kou'an" with treaty ports, but also residence and trade. Reminding the government that its registration petitions never used the term foreign concession, Jiuda as a Chinese company should not be restricted by a designation applicable only to foreign-owned entities. Citing several salt refineries operating inside Shanghai's foreign concessions without the Salt Administration's approval, Jiuda need not have bothered with registering (and subject itself to all the regulations) with the government had its operation been confined to the concessions. Indeed, its Tianjin and Hankou outlets, both approved by the Salt Administration, were located in the Chinese part of the cities. It also opposed the Salt Administration's permission for Yihexiang to verify refined salt shipment: that policing authority belonged to the state, not a private party, much less a revenue farmer predisposed to hostility against Jiuda. It finally filed a suit with the Administrative Court on December 15, 1919, against the Salt Administration, seeking to vacate order #520 and the restoration of the status quo ante of July 1918 under order #939.

The Salt Administration's April 1920 brief defended order #520. While foreign concession and "commercial port" were indeed different terms, they overlapped in defining the boundary of foreign trade and residence. As such, the Salt Administration was merely exercising its discretionary authority in regulating the salt trade, including those Chinese-owned refineries operating inside foreign concessions. Jiuda's mission of competing against a tax-free import and improved quality of salt still enjoyed the full support of the government. On the other hand, until the revenue farming regulations were repealed, they must also be enforced. Chiding both Jiuda and the revenue farmers for their selfish interests, the government was striving for a balanced solution. The order was thus applicable only to Nanjing and not retroactive, leaving Jiuda's preexisting outlets intact.[23]

Meanwhile, an equally dissatisfied Yihexiang also sued the Salt Administration on January 4, 1920. Citing the Foreign Ministry's first recommendation, the suit claimed that Nanjing city should be classified as "interior" (*neidi*) and therefore not part of the port. Permitting Jiuda's access was thus not merely a violation of settled gabelle regulations but also opened the door for foreigners to infringe further upon China's sovereignty. Accordingly, it sought relief by the Administrative Court to vacate order #939 of July 1918.

The Salt Administration's brief against Yihexiang's suit emphasized the Foreign Ministry's second reply, which affirmed the administration's discretionary authority to adopt a broad definition of "tongshang kou'an." Confining Jiuda's outlet and operation to only the designated area to combat a foreign import already protected Yihexiang's privileged status. Giving residents inside that area a choice did not compromise the plaintiff's privilege, nor would its order invite foreign business to operate inside Nanjing city. Jiuda was a Chinese company and what rights (if any) it enjoyed could not be extended to nor invoked by foreigners. The state's interests in salt were threefold: consistent application of the gabelle rate, consumer welfare, and choice. In this case, Jiuda was paying the same rate as Yihexiang while offering a superior product to the consumer at a higher price—none of which, in the government's view, affected the plaintiff's interests.[24]

On January 26, 1920, the Administrative Court invited Jiuda to brief the court in Yihexiang's suit against the Salt Administration. Jiuda took the opportunity to fault both the plaintiff and the defendant. Reiterating the arguments in its own separate suit, the company added that, in determining the boundary of "tongshang kou'an," the residence test—whether foreigners were explicitly permitted by the Chinese government or by treaty—was a moot one. Even by the Foreign Ministry's own admission, plenty of foreigners already resided in Nanjing outside of Xiaguan. In Beijing, declared by the Salt Administration as a noncommercial port and therefore closed to Jiuda, imports filled the shelves of Chinese and foreign-owned emporiums.²⁵ Neither foreign residence nor trade thus observed any Chinese administrative boundary. At any rate, referring again to the company's registration petitions, Jiuda was not founded to focus exclusively on competition against foreign imports. It was also established to improve the quality of salt produced in the country.²⁶

Packaging

While Jiuda and Yihexiang battled over the definitions of "commercial ports," Gonghengmao et al., representing the Lianghuai revenue farmers and producers, also filed a separate suit with the Administrative Court. They argued that the Salt Administration's order of December 9, 1919, must be vacated as it had been negligent in investigating the amount of foreign salt imported and thus failed to limit the domestic production and sales of refined salt. In addition, since Jiuda's first registration petition had specified its product as "powdered salt in paper packages" (*zhibao fenyan*), packaging should also be an issue. Any other packaging method, including use of gunny bags, thus violated the terms of its registration. Citing longstanding regulations that any change in packaging material must be authorized explicitly and the shipping weight of salt verified by the state, the plaintiff charged that the Salt Administration was inconsistent in enforcing its rules and negligent by creating opportunities for smuggling. For reasons unknown, Jiuda was not asked to brief the court in this case.

Responding to Gonghengmao et al., the Salt Administration emphasized that production and marketing of refined salt was authorized by the Special Regulations on Salt Production. As a registered company, Jiuda was subject to those stringent rules, including limitations on both its production and marketing. The government, in fact, did monitor the content and weight of Jiuda's shipment, regardless of the packaging, whether in bottle, paper, cotton cloth, or gunny bag. To support its accusation, the plaintiff must provide evidence of smuggling or overpacked bags. Indeed, it was the plaintiffs' failure to improve the quality of their products that led to rampant smuggling of foreign salt (*yangsi*), a severe problem about which the Maritime Customs obviously could not furnish any reliable statistics.²⁷

On August 17, 1920, the Administrative Court ruled against both Jiuda and Yihexiang, upholding the Salt Administration's order #520. It was a question of intent. Sidestepping the Gordian knot of defining "tongshang kou'an," the court focused on Jiuda's declared intention to compete against tax-free imported salt. The court thus found the Salt Administration's order consistent with the Foreign Ministry's recommendation that there must be a spatial limit, whether broad or narrow, on the right to trade.[28] Four days later, it ruled in favor of the plaintiff in *Gonghengmao et al. vs. the Salt Administration*. With Jiuda not a party to the proceedings, the court overturned the December 1919 order and, by extension, ruled against the company. Ignoring the problem of smuggling, the court found that the government had failed to exercise due diligence in determining the amount of salt imported. As foreigners declared only 288 *dan* of salt imported through Shanghai in 1918, a record high as reported by the Maritime Customs, Jiuda's production and actual shipments had obviously far exceeded that number. Given the company's declared intent, Jiuda had more than achieved its goal. The court instructed the Salt Administration to conduct a thorough investigation to determine the amount of imported salt and limit accordingly the production and sales of refined salt. As for packaging, the company should follow what was specified in the "original" case (*yuan'an banli*).[29]

The verdicts came as a surprise. Recalling the episode many years later, Jing Benbai traced the defeats to a negligent, if not corrupted, Administrative Court and Wu Fengqiao, chief of the Production Bureau at the Salt Administration who was hostile to the refined salt cause.[30] What Jing omitted in his account was that both Li Sihao (director-general from December 6, 1919 until July 29, 1920) and Zhang Hu (chief commissioner from January 10, 1919, until August 18, 1920) had left office in another political reshuffle. In a panic, Jing wrote to Fan Xudong urging him to implore Yuanlian, again appointed acting Education Minister on August 11, 1920, for the assistance of Zhou Ziqi (Finance Minister) and Pan Fu (chief commissioner, August 18, 1920 to November 5, 1921) in reversing the decisions, and to find someone who could approach various provincial authorities in forming a new company to overcome the revenue farmers' resistance.[31]

A more collected Jing soon resumed the campaign through a series of petitions to the Salt Administration. The court's reasoning had ignored the company's declared goal of producing a modern hygienic product. Furthermore, the company's first application for registration had not received official approval, and the final petition of September 1914, which won approval, did not specify the packaging method. Pointedly, he questioned whether such specificity was necessary. For instance, none of the company's registration petitions made any reference to the mode of transportation. By the court's reasoning—what was not specified in the document would be illegal—the company could not ship its product! He urged the Salt Administration to clarify the matter with the court.

Help for Jiuda came from an unexpected source. Following the Administrative Court's decision, the Changlu commissioner began to withhold Jiuda's shipping permits.[32] Alarmed by the sudden drop in revenue and not unsympathetic to the company's plight, Sir Reginald Gamble offered to bypass his Chinese colleagues and issue the permits on his authority.[33] Circumventing the court's decision, Jiuda, the Changlu Inspectorate and the Chief Inspectorate agreed on a "new" packaging method: powdered salt packed in paper inside gunny bags (*madainei baozhi fenyan*). Following elaborate maneuvering behind the scene, Pan Fu, confronting a fait accompli and unwilling to compromise his authority of issuing shipping permits, finally telephoned Ding Naiyang, the Changlu commissioner, authorizing the release of Jiuda's permits with the new packaging.[34]

Opposition, however, continued. On June 2, 1921, the Salt Administration's Songjiang suboffice seized two Jiuda shipments of "refined salt packed in paper inside gunny bags" totaling sixty tons for violating the Administrative Court's August 21, 1920, ruling.[35] The company's petitions for immediate release of the seized shipments were rejected by Wu Fengqiao. The shipping permits might be genuine, but there was no written record that his office had approved the new packaging method (Pan had authorized the arrangement by phone), and until the Administrative Court's ruling was vacated, refined salt must be packed and shipped only in paper bag. Left with no alternative, Jiuda requested the Administrative Court to clarify whether "powdered salt packed in paper inside gunny bags" would comply with its decision. In two terse sentences, the court replied that the packaging method, as per the shipping permit issued by the Changlu Salt Administration, was beyond the scope of its August 21, 1920 decision. However, it declined to instruct the Salt Administration to release the seized shipments. That authority belonged to the Salt Administration.[36]

With the ball back in its court, the Salt Administration, uncertain of the meaning of the Administrative Court's terse findings, queried again on July 27, 1921. A somewhat irritated court replied that its ruling was clear: Jiuda's "original" petition should be binding. Any change in packaging would deviate from the "original" and therefore violate the decision. However, as the seized "refined salt packed in paper inside gunny bags" appeared to be transported under genuine permits issued by the Salt Administration, it could be a new fact (*xin shishi*) and therefore outside the scope of its decision. How that permit was issued fell under the Salt Administration's jurisdiction. Not even ex-president (and shareholder) Li Yuanhong's personal appeal for support (*weichi*) of the company could dissuade Pan Fu from upholding the rule of law.[37]

Indeed, the Administrative Court's clarification gave Jiuda's opponents in the Salt Administration more ammunition. When the Changlu commissioner reported on September 8, 1921, that Jiuda's shipping permit of "powdered salt packed in paper inside gunny bags" had been authorized by phone, Beijing faulted it for procedural violation: an order must be written to be legally valid. As Ding Naiyang had been dismissed and Pan Fu struggling (he was removed on November 5, 1921), Jiuda was left holding the bag. So long as gunny bags were used in shipping refined salt, the company was violating the court's decision. Furthermore, in Wu Fengqiao's interpretation, a "new fact" meant that the substance being shipped, not merely the packaging method, must also be "new."[38]

An exasperated Jiuda fought back by exploiting the divisions within the salt bureaucracy. While filing more petitions challenging the Salt Administration to design a paper packaging that could survive the journey from Tanggu to Central China, the company also reminded the Revenue Inspectorate that the Changlu commissioner had issued the shipping permits. Could the Songjiang suboffice overrule a superior office, and if so, it demanded a refund of the gabelle.[39] Before going on leave, Gamble faulted his Chinese colleagues for not briefing the Administrative Court properly, resulting in such troubling decisions. W. Strickland, acting in Gamble's absence, was equally frustrated with this tempest in a tea pot. Even if Jiuda's first petition for registration contained the Chinese words "packaged in paper" (curiously omitted in the English translation, raising the issue of which version would be governing, an issue he did not pursue), none of the six subsequent petitions until the company was finally approved for registration specified the packaging material or method to be used. If the company could (and did) revise its registration petition, he asked, why should it not be permitted to do so now without being deregistered? If, indeed, the Revenue Inspectorate had been appraised of and approved the use of different packaging materials and methods as early as 1917, did those precedents render the Administrative Court's decision moot?[40] Above all, gabelle revenue would, in his opinion, certainly suffer, and the servicing of various loans and bonds affected. However, Pan Fu's successor, Zhong Shiming (1879–1965), a Harvard graduate in government science, declared his hands tied. He reminded Strickland of the limits to the inspectorate's authority. Even the Finance Minister, whom the inspectorate served, must abide by the Administrative Court's decision. By law, the Administrative Court's ruling was final.[41]

Jiuda's solution to the conundrum appeared on November 18, 1921. Late that day, Jing alerted Fan to a golden opportunity. Politics had engulfed the banks, leaving the Beijing government without cash to pay for the capital garrison. With a mutiny threatening, President Xu Shichang ordered Zhong Shiming to raise 300,000 yuan by noon the next day. With his effort falling short, Zhong approached Jing for a 80,000 yuan bridge loan (if more could be raised, even better), implying that issues such as packaging, gabelle rebate, and production cap were all negotiable. Four telephone calls between Beijing and Tianjin later, the amount was remitted. Building on the goodwill, Zhong met and discussed candidly with Jing his disagreements with Zhang Hu and Li Sihao, but he, too, as a Western-educated official (and a member of the Young Men's Christian Association), shared the goals of gabelle reform and breaking the revenue farmers' hold on the country. Briefed by Jing on a possible solution to the packaging problem, Zhong instructed Wu Fengqiao to be as accommodating as possible (*shefa tongrong hezhun*).[42] After more word-mincing, the result was another new Jiuda product: refined crystal salt (*li jingyan*) shipped in gunny bags.[43] Queried again, the Administrative Court ruled the product as a new fact in both substance and procedure. Although the Songjiang suboffice again seized a Jiuda shipment while the revenue farmers lobbied for a Salt Administration order to stop all further shipments, a personal appeal from Liang Qichao to Sun Muhan, then Commissioner for Taxation, stayed the order and bought more time for the company.[44] Finally, the Salt Administration circulated order #600: "refined crystal salt in gunny bags" did not contradict the Administrative Court's August 21, 1920 decision.[45] The problem finally behind it, the company continued to grow.

Production Limit

Still, a production cap limited Jiuda on how much it could legally produce. To reduce cost and exercise more quality control over the supply of raw material, Fan Xudong purchased three salt pans in

1918.⁴⁶ Separate from Jiuda, they were registered in Fan's name as his personal property. He became a contractor to the company, thus circumventing another longstanding principle of salt administration: segregation of salt producers from distributors (*chan bu jian yun*) for fear of collusion and smuggling between the two. With its own depot supervised by the Changlu Inspectorate and a new plant capable of producing 180,000 *dan* a year coming on line, Jiuda applied to remove its production cap. Whereas the refining process from crude salt could be limited by regulating purchases from the state depot, nature determined the size of the harvest from the evaporation ponds. Jiuda enjoyed a standing order from International Exports amounting to over 80,000 *dan* a year, and more revenue could be generated for the state. Rejected initially, the application was finally approved because the Salt Administration was concerned by Japanese smuggling in the Northeast. In exchange for Jiuda's expertise to construct a new state-owned refinery in Yingkou (which was never built), the company's annual production cap was raised to 300,000 *dan*.[47]

In 1920, following the opening of Western Hubei to "free" trade in salt and inquiries from Vladivostok and Southeast Asia, Jiuda sought to double the company's annual total output to 600,000 *dan*.[48] By then, however, the revenue farmers' suits and shifting Beijing politics had taken their toll on the company's plan. Citing the Administrative Court's decisions, the Salt Administration denied the application. The shipment to Vladivostok in 1919 was a *pro hac vice* special case, demand from Southeast Asia sporadic, and access to Western Hubei free only for crude salt. Indeed, with new refineries appearing (see next chapter), the government was already concerned with possible excess capacity. Lifting Jiuda's production cap yet again was out of question.[49]

Mounting a petition and cable campaign, Jiuda spent the next two years solving the problems of market access and production limit.[50] As a result of the lawsuits, in January 1923 the salt bureaucracy finally called a joint session—unprecedented, according to Jing Benbai—to address the marketing problems caused by refined salt. Three representatives each from the Revenue Inspectorate and the Salt Administration met in secrecy for two months drafting the "New regulations on company registration, transportation, distribution, and gabelle payment for refined salt." Total annual production for refined salt was capped at 2.5 million *dan*. A free trade surcharge of 0.5 yuan/*dan* on the refineries gave them access to all divisions in the country. As a placating measure, however, the surcharge would be waived if the refined salt was distributed by existing revenue farmers.[51] Although these measures were never implemented as a result of lobbying by the revenue farmers, the Salt Administration queried the Administrative Court as to whether the new regulations might supersede its decision of August 21, 1920. With an affirmative answer from the court, the Salt Administration raised Jiuda's annual production cap by another 10,000 tons to a total of 450,000 *dan*. Whether the court's decision was facilitated by a loan of 60,000 yuan from Yongli, Jiuda's sister company, to tide the court over the mid-autumn festival, or influenced by Zhang Hu's brief return to office as Finance Minister and concurrently director-general of the Salt Administration from August 14, 1923, to November 12, 1923, remains a matter of conjecture.[52]

Consumers' Right to Choose

We do know that political influence and bribery helped Jiuda clear its last unresolved legal hurdle: the permission for consumers residing beyond the commercial ports (however defined) to purchase, transport, and consume refined salt. Long the cause of many lawsuits, persecutions, and confiscations, the issue not merely pitted the revenue farmers against Jiuda, but also the Salt Administration in Beijing against the provinces. Protecting their exclusive privilege in supplying crude salt to their designated districts, revenue farmers considered refined salt outside the commercial ports a smuggled contraband and, with the cooperation of provincial salt bureaus and revenue guards, routinely confiscated such shipments, fined shippers, and prosecuted owners as smugglers. Jiuda and its dealers, however, contested the issue with the provincial governments and Beijing: how could refined salt, with gabelle fully paid to the central government and transported under shipping permits issued by the Salt Administration, be considered a smuggled contraband? Could consumers in the interior be denied the choice, if not the right, to consume the kind of salt they want?

Jiuda's opportunity came with Li Yuanhong's return to power as president of the republic in June 1922. A revitalized gabelle reform movement introduced no less than thirteen resolutions when the National Parliament reconvened. The Hubei delegation, supported by petitions of Hubei sojourners in Beijing, proposed the abolition of revenue farming and the institution of free trade of salt for the province.[53] With minimal lobbying by Jiuda, the Hubei provincial assembly also passed a resolution on June 8, 1923, legalizing the purchase, transportation, and consumption of refined salt beyond Hankou, Shashi, and Yichang. The people should be free to choose a hygienic product for their own health, the argument went.[54] On the other hand, while conceding that consumers in the interior could purchase unspecified quantities of refined salt for incidental consumption (*lingxing goushi*), the Salt Administration insisted that there must be no trade (*fanyun*). A defiant provincial assembly adopted a resolution on July 11, 1923, rejecting Beijing's argument. Denial of trade amounted to a de facto ban on consumption, and so long as the gabelle was paid in full, residents of the province should be free to choose between crude and refined salt.

Xiao Yaonan the military governor was caught not merely between the Salt Administration and the Provincial Assembly. A letter from Li Yuanhong's personal secretary obliged him to "support strenuously" (*dingli weichi*) Jiuda.[55] Although Li was toppled in a coup engineered by Cao Kun on June 13, 1923, Cao, too, was "deeply interested" in Jiuda as a shareholder.[56] On the other hand, there were also the provincial salt bureau and revenue farmers reportedly with a 200,000 *tael* lobbying war chest to deal with.[57]

To end the tug of war, Xiao Yaonan tried to find a mutually acceptable solution. The bureau, citing established antismuggling regulations, allowed no more than ten catties per person. Any more would jeopardize antismuggling enforcement and erode sales of crude salt, with an adverse effect on its performance review. Jiuda and representatives of various chambers of commerce countered with 100 bags (approximately six tons), insisting that a distinction be made between tax-free smuggled salt and refined salt with gabelle paid in full. Whether Jiuda's gift of 14,500 yuan in Yongli shares and payment of 20,000 yuan to Xiao Yaonan's chief-of-staff reinforced the governor's decision is uncertain, but Xiao did propose a twenty bag per person per purchase rule to the Salt Administration as a solution.[58]

Beijing again became the focus of intense lobbying by both sides. The provincial salt bureau chief's report echoed the arguments of Lianghuai's revenue farmers that it was impossible to distinguish between "trading" and "consumption" in antismuggling operations and that the proposed twenty bag rule was excessively generous. With an average consumption between eight and twelve catties per year per person, no individual would purchase almost 3,000 catties of refined salt with the intention of consuming it all alone. Furthermore, what was to prevent shipments of 1,000 bags (approximately sixty tons) under a bundled purchase of fifty "consumers"? The effect of such incidental consumption would be a blatant violation of existing gabelle rules and a disaster for the Lianghuai division.[59] Countering this effort, Jiuda also mobilized its network. Ye Kuichu, a shareholder, was asked to discuss the matter with his friend Qian Fangshi, another Harvard graduate and head of the Salt Administration (August 18, 1923 to October 27, 1923). Jing Benbai did his part by keeping Zhang Hu informed, once tracking the vacationing minister of finance to Tantou Temple in the western suburbs of Beijing.[60] Although Zhang again left office on November 12, 1923, the Provincial Assembly kept the pressure on Xiao Yaonan, promulgating the twenty bag rule as law and ordering the Salt Bureau to stop interfering with any refined salt purchase and shipment under that limit.[61] Access to the Hubei market was finally settled and a precedent set for other provinces such as Hunan. Jiuda prospered with access to the market. By 1924, after a stock-split and dividends, early investors in the company had more than recouped their original investment, each share fetching as much as 165 yuan (par value 100 yuan) in Tianjin's stock exchange. The company raised capital and loans from the market with ease, a far cry from its early days (see Appendixes 2 and 5 for the company's profits, dividends, and shipment destinations).

As reconstructed above, the saga of how Jiuda broke into the country's highly regulated salt market illustrates the efficacy of its network strategy. Drawing on the resources of its shareholders and associates—Li Yuanhong, Cao Kun, Chen Guangyuan, Liang Qichao, Fan Yuanlian, Zhang Hu, Li Sihao, and Sun Muhan, to name but a few—the company overcame various hurdles along the way.[62] From creating a national marketing system to lobbying the Salt Administration on diverse policy issues, the strong ties embedded in its shareholders' network played a crucial role in the company's survival.[63] Even if access (*menlu*) to critical officials or institutions could not be found through its network, Jiuda manufactured relationships by paying to play, as in the case with Zhong Shiming, or gifts to foreign staff members of the Revenue Inspectorate, or shares to Cao Kun's key aides, or loans to the Administrative Court.[64] To level the playing field against revenue farmers employing similar practices, it offered gifts and bribes to incorporate influential officials such as Xiao Yaonan and his chief-of-staff into its shareholder network, or cash to the commander of the Hubei provincial antismuggling force.[65] As forms of corruption, perhaps a distinction may be made between rent creation (win-win) and rent extraction or political extortion (zero-sum). While costly, it was instrumental in Jiuda's attempt to enter the salt market by attacking the regulatory regime of revenue farming, a process of creative destruction identified by Joseph Schumpeter as the essential fact about capitalism.[66] Whether the process was sustainable and the price paid by the company for its success we shall see in the next chapter.

4

The Price of Success

Once Jiuda breached the barriers of entry and reaped the profits of institutional innovation, other refineries soon appeared. Although not every attempt was successful, the industry grew. But while Fan Xudong and his colleagues celebrated the development as an encouraging sign of the country's improving hygiene and free salt trade, they soon found Jiuda paying the price of success. As the country continued its descent into political disintegration—even the once powerful Revenue Inspectorate often rendered helpless by provincial militarists—Jiuda, the largest of the refineries and therefore the most conspicuous target for rent seekers, was forced to shut down several times. Even worse, the dark side of network strategy began to haunt the company. Should Jiuda continue the creative destruction of the revenue farming system and promote the growth of competing modern refineries? This chapter reconstructs how, in the face of mounting competition from other refineries, Fan and his colleagues recalibrated their strategy of creative destruction to collaborate with select revenue farmers and worked against competing refineries to preserve Jiuda's share of the refined salt market.[1] It was a small price to pay for its patriotic mission to developing the country's chemical industry as a "Chinese" elder brother.

A Growing Industry

Carrying out their threat to crush Jiuda, the Lianghuai revenue farmers launched the Lequn Salt Refinery in 1919 at Yangzhou with a capital of 300,000 yuan. However, for reasons unknown (although Jing attributed that to ineptitude), the company failed to begin regular production and its registration revoked in 1924.[2] Similarly, the

Changlu revenue farmers, led by Li Baoxian and other head merchants, applied for the registration of Xianfu Salt Refinery at Han'gu, Ninghe district, in Hebei. Capitalized at 100,000 yuan, the proposed company was to serve commercial ports and any other ports declared open for trade with an annual production capacity of 100,000 *dan*. In addition, the application specified that a refund was due when it shipped to divisions with a gabelle rate lower than Changlu. The head merchants obviously felt, as did Jing and Fan earlier, that the refund was justified, but they certainly overlooked Jiuda's experience with the Revenue Inspectorate. Already embroiled in litigation over the definition of "tongshang kou'an" while defending its revenue principle, the Salt Administration rejected the application.[3] After further revisions, the company's application finally received approval with no mention of any gabelle refund, although it, too, never began regular production.

In contrast, Tongyi (Tung I) Salt Refinery, Ltd., founded in 1919, presented a far more serious challenge to Jiuda. Headquartered also in Tianjin but with its production facility located at Yantai, Shandong, the company was part of Tunway Enterprises, one of modern China's earliest holding companies controlled by the Sun family of Anhui.[4] Descendants of Sun Jia'nai (1827–1909), a grand secretary and tutor to the Guangxu emperor, this prominent household straddled the worlds of officialdom and industry. In a household division of labor, several members of the family served as high officials, while others crisscrossed the modern-traditional barrier as revenue farmers in Yangzhou and as managers of industrial enterprises, including Qixin (Chee Hsin) Cement Co., Ltd., the Kailian (Kailan) Mining Administration, flour mills in Shanghai, Xinxiang (Henan), Ji'ning (Shandong), and Harbin with Zhongfu Bank (Chung Foo Union, founded 1916 and headquartered in Tianjin until 1930 when it moved to Shanghai) at the core.[5] Tongyi, initially capitalized at 250,000 yuan, enjoyed the resources of not merely its own house bank but also the Sun's vast network, including Gong Xinzhan (1871–1943; assistant finance minister, 1916; finance minister and acting prime minister, 1919; interior minister, 1924–25; and transportation minister, 1926) as the company's chairperson, and other prominent board members or shareholders such as Fu Zengxiang (1872–1950; education minister, 1918–19), Xu Shizhang (1886–1954; younger brother of President Xu Shichang; vice-minister, Ministry of Communications, and vice-president of the Bank of Communications, 1920–1922), and Nie Qiwei (industrialist and a grandson of Zeng Guofan).[6] With an initial annual capacity of

60,000 *dan* (reaching 400,000 *dan* by 1932), Tongyi was more than a match for Jiuda in financial and network strength.

Equally potent competition came from Wuhe, another limited liability company established in 1928 in Shanghai with an initial capital of 100,000 yuan by several leading Zhejiang revenue farmers, including Zhou Qingyun (1864–1933) and Wang Shoushan (1873–1938). Other major shareholders included Zhang Zhanru (fifty shares), an old friend of Jiang Jieshi and brother of Zhang Renjie (one of the four elder statesmen of the Nationalist Party), as well as the salt merchants' guilds of the five prefectures of Jiangsu (*Suwushu*, including Suzhou, Songjiang, Changzhou, Zhenjiang, and Taicang) and Liangzhe (Yuyao-Daishan-Songjiang gongsuo).[7] Jiuda's intelligence sources reported that while the company was well-funded (Zhou, Wang, and Zhang had already invested 160,000 yuan into the enterprise), its production was geared toward washed salt, a slight improvement over crude salt but nonetheless still unrefined.[8]

In addition to revenue farmers, investors from a variety of backgrounds all over the country also entered the industry. Minsheng Refined Salt Co., Ltd. of Dinghai, founded in 1927 by a group of Zhejiang investors (including Fang Xiangeng, a charter Jiuda shareholder) with a capital of 50,000 yuan, shared Jiuda's mission of improving the quality of salt consumed in the country. Like Jiuda, it had trouble persuading the Salt Administration on the use of gunny bags for its shipments. For over two years, the company filed petitions after petitions for registration until it finally received approval in 1929.[9] At least two other applications were denied because the proposed manufacturing sites were within one hundred *li* of Tanggu, a violation of Jiuda's geographical monopoly as granted by the Salt Administration.[10] Entry barrier into the business remained high.

Nevertheless, by 1928 the refined salt industry was a far cry from Jiuda's small beginning. Excluding defunct ones, the paid-up capital of registered and operational salt refineries had grown from 50,000 yuan in 1914 to over 4 million. Increase in production capacity mirrored the growth in capital. From the 30,000 *dan* in 1915, production capacity of registered salt refineries by 1919 had reached 320,000 *dan*, a tenfold increase. Between 1920 and 1926, the pace of growth slowed, although the 1.43 million *dan* increase in production capacity far exceeded that of the previous period. Thereafter, production capacity grew by another 1.1 million *dan*, or 77 percent. In terms of capital and production capacity, unprecedented growth for the refined salt industry thus took place between 1920 and 1926.[11] Indeed, had the government been less stringent, there could have been at least forty more refineries (see Appendix 6 for a list of registered refineries in 1935).[12]

Jiuda's Strategy

As other salt refineries joined the industry, Jiuda responded by raising more capital to finance its expansion through vertical and horizontal integration, securing monopoly privileges, collaborating with revenue farmers in Shandong and Shanghai, and adopting restrictive practices, including informing the authorities on its competing refineries. The company's paid-up of capital doubled to 300,000 yuan in 1918 and 500,000 the following year (see Appendix 1). The shareholders' meeting of 1920 discussed a bond issue of 300,000 yuan. Much of the company's 500,000 yuan paid-up capital had gone into physical plant and equipment. At least another 700,000 yuan in liquidity was needed to finance the gabelle advance and operations. Generous credit lines from the house banks had covered the shortfall but also reduced the company's profitability. Offered at an annual interest rate of 8 percent and secured by all the assets of Jiuda, the bond, with one-tenth callable each year, would be repaid in ten years.[13] After further deliberations, the board decided that the company needed even more to finance the development of the country's chemical industry. With a record of profitability, the 1921 capital campaign enjoyed an enthusiastic response, a far cry from its early struggle. Oversubscription forced the company to increase its total paid-up capitalization from a targeted 1.6 million yuan to 1.7 million. In 1923, another annual shareholders' meeting authorized a rights issue of 400,000 yuan to finance Yongyu (see next section), bringing total capitalization of Jiuda to 2.1 million yuan.[14]

Such easy money financed not merely Jiuda's expansion but also realization of Fan Xudong's dream of developing the country's chemical industry with the launching of Yongli Soda Ash Co. (also known as the Pacific Alkali Co., Ltd.) in 1917 to manufacture sodium carbonate. Residuals from Jiuda's refining process became raw material for a by-product plant to produce tooth paste and powder, mouth rinse, and magnesium carbonate in 1918. For research and development, Fan and his colleagues endowed the Golden Sea Chemical Industry Research Institute headed by a Harvard-trained chemist Sun Xuewu (1888–1952) in 1922. When part of Tanggu's wharf, hitherto under Russian control, became available, Jiuda purchased it through Hengfengtang, its holding corporation, and eventually expanded into shipping as a major shareholder of the Tianjin hangye gongsi (Tientsin S.N. Co., Ltd.) in 1929.[15]

Partnership with Revenue Farmers

As much as Fan and his colleagues wanted to replace the revenue farming system with free trade, their creative destruction proved difficult to sustain. Jiuda's success created vested interests, including the need to protect its share of the refined salt market. Collaboration with revenue farmers in select markets also made good strategy if only to understand its enemy better. Under German control before the First World War and the following Japanese occupation, Qingdao had developed a sizable salt (both crude and refined) export industry. In 1922, the Beijing government insisted that no foreigners be allowed to produce salt in China, and began negotiations with Tokyo over the salt flats and refineries. Japanese owners, supported by their government, demanded 30 million yuan for their investment while retaining the export right to Japan and sell any surplus in China. To Jing Benbai, that was a major threat. As an insider, he was privy to a costly Salt Administration survey that identified Tanggu and Qingdao as two ideal sites for salt production. Tanggu enjoyed sea water of high salinity, long dry summers, a cheap source of coal from the nearby Kailan collieries for fuel, easy access to rail and water transportation, and an abundant supply of cheap land along the long and flat coast line. Qingdao enjoyed all that and more: a deep sea harbor convenient not merely for exports but also, compared to Tanggu, half the shipping cost to the Yangzi provinces. In the wrong hands, Jiuda would have a formidable competitor.[16] Somehow, Jiuda needed to turn the threat into an opportunity.

The company's reputation served it well. Recognized for his expertise, Jing Benbai was appointed an advisor to the government in the negotiations.[17] He recommended Yang Zinan, a Jiuda engineer, to appraise the Japanese properties. To avoid a bruising bidding war, Jing negotiated a partnership with the revenue farmers of the Shandong division, once the company's nemesis. Together, they formed the Yongyu (Yung Yu) Salt Co., Ltd. (capitalized at 3.2 million yuan, but only 800,000 paid up, of which Jiuda contributed 250,000) to enter a bid. After disqualification of a 5.4 million yuan offer from a nominee for Japanese interests, Yongyu won with a 3 million yuan bid, payable by annual installments over fifteen years to the Salt Administration. Under Fan's general management, Jiuda controlled the refining operation while the revenue farmers handled the production and export of crude salt to Japan.[18] Jincheng, Jiuda's house bank, generously provided an 800,000 yuan line of credit at 12 percent monthly interest to the new company.[19] Although the acquisition caused Fan Xudong and Jiuda much grief, as we shall see, the company nevertheless strengthened its position as the refined salt industry's leader in both capital and production capacity.

Jiuda's partnership with revenue farmers of Shanghai also offered a variety of benefits. Since 1917, Gongmao, partly owned by Zhou Qingyun (major shareholder, too, of Wuhe) and the Jiangsu Five prefectures guild, had served as the exclusive supplier of crude salt to the Shanghai International Concessions, paying 190,000 yuan annually for the monopoly. When that privilege came up for renewal in 1927, Jiuda joined the bidding. Through Hengfengtang, its holding corporation to mask its involvement, the company invested 45,000 yuan in Jiuhe, a partnership with two leading Shanghai capitalists: Shen Renfu, manager of Ningbo Commercial (Siming) Bank, and Yu Qiaqing (1867–1945). Offering 423,000 yuan per year on a volume of 283,000 *dan* and 100,000 yuan payable immediately as security, Jiuhe won. However, Gongmao refused to relinquish the monopoly by advancing 200,000 yuan to the cash-starved Salt Administration. Officials ignored Jiuhe's protests and the impasse was resolved only after officials brokered a deal in August 1928: Gongmao and Jiuhe formed a new partnership to operate the Shanghai monopoly.[20]

Existing archival material is silent on Fan Xudong and Jing Benbai's motives in these decisions. Perhaps their quarrel had been more against the system than the revenue farmers themselves. Collaboration with Shandong revenue farmers avoided a costly bidding war, and allowed Jiuda to retain control of the production of refined salt. Even when offered a good price, Fan declined to sell his share of the Jiuhe partnership. Discussing possible Jiuhe's investment in Yongyu, Zhou Qingyun acknowledged Fan as a man of integrity.[21] Collaboration, as a matter of business strategy, also gave Jiuda insights into the opposition and the country's soy sauce market, which eventually led to founding of China Chemical Industries (Quanhua huaxue gongye gongshi) in 1935 for the manufacture of soy sauce in Nanjing.

Against Competing Refineries

If foes became partners, competition also drove Jiuda against other refineries. In 1921, Tongyi broke into the market with an aggressive strategy: an interest-free month long grace period for deliveries made, accepting remittance drafts issued by local native banks as payment (considered a risk with additional handling charges), offering rebates, and packing extra salt free of charge. Matching such practices proved costly and effectiveness only temporary.[22]

When competition failed, Jiuda resorted to a somewhat underhanded strategy—informing on its rivals. Xu Zhaozhou, Fan's relative by marriage and manager of the company's Hankou office, sent an anonymous letter alerting the Maritime Customs Service to Tongyi's overpacking and underreporting of its shipments.[23] When the Customs failed to act on the tip, allegedly because of its close relationship with Tongyi's manager, Jiuda went public in local newspapers. The campaign was costly: to Xu's conscience and in payoffs to publishers.[24] Similarly, Fan Xudong complained to Miao Qiujie (1889–1966, Dane's protégé, then Lianghuai salt commissioner) that Fengtian, another refinery, had been packing three to six extra pounds of salt into each bag of salt.[25]

Partnerships with once loathed revenue farmers and informing on competing refineries, Jiuda paid the price for its success in the form of conscience attack, its principle of free trade, and bribes to newspapers. There was no permanent enemy in business, or friends in politics.

Political Disintegration

Indeed, if Jiuda's shareholder network, counting among it influential national and provincial political figures, contributed to the company's success, the tumultuous republican period also exposed the limits, risks, and uncertainties of networking as a business strategy as the company's network was neutralized in Hunan, ignored in Jiangxi, overwhelmed in Hubei and Anhui, or betrayed in Changlu.

In July 1920, Tan Yankai (1879–1930), who regained power a short while ago as military governor of Hunan, imposed a levy on refined salt. Lobbying by prominent Hunanese officials, including Xiong Xiling, Guo Zongxi (governor of Ji'lin), Chen Jie (Commercial Bureau chief, Ministry of Industry and Commerce), Liu Difan (Hunan provincial financial bureau chief, 1913), Liu Kuiyi, and Li Binshi, the last two also among Jiuda's earliest shareholders failed to dissuade Tan. His terse reply to the lobbying notables insisted that the levy was only an additional 0.05 cents per *dan*, the rest of the two yuan provincial levy merely legitimizing preexisting customary fees and a military surcharge.[26] Perhaps an old tax is a good tax, and Tan, as a leading statesman and the principal graduate of his class in the Metropolitan Examination of 1904, certainly could match the lobbyists' influence and prestige. Table 4.1 summarizes, according to company records, the extractions from Jiuda by various provincial authorities between 1921 and 1925.[27]

The company's franchisee was forced to close down at least once.[28] Nevertheless, with these demands becoming routine and secured against future tax receipts, Jiuda found them a small price to pay.[29]

But the price could be beyond the means of the company, as in Jiangxi from 1917 to 1926. As a shareholder, Chen Guangyuan (1873–1939), the military governor, had granted the company access to his province over the opposition of revenue farmers and the provincial salt bureau.[30] The company's franchisee, Wu Jinyin, brother of the Northern Jiangxi garrison commander, arranged an additional levy of 0.07 cents per *jin* of refined salt in exchange for access to the interior via Jiujiang and Nanchang.[31] Business boomed.

However, shifting politics also left the company vulnerable. After Cao Kun, another shareholder, ousted Chen from office in 1922, Jiuda's business in the province entered a period of uncertainty. Cai Chengxun, Chen's successor, seized on salt as a quick way to raise revenue while amassing a personal fortune of over 1.7 million yuan in his three-year reign.[32] In 1925, politics in Jiangxi underwent another drastic realignment after the fall of Cao Kun. Fang Benren became the provincial military governor, and his cousin Fang Zhaode the salt commissioner. Exploiting the change, the revenue farmers advanced a loan of 600,000 yuan in exchange for a ban on refined salt. Playing both sides, Fang demanded a contribution of 2 million yuan from the refineries for him to lift the ban.[33] Wu Jinbiao's attempt to intercede

Table 4.1 Extractions from Jiuda by the Hunan provincial government, 1921–1925

Month/Year	Amount (yuan)	Reason
12/1921	10,000	1,200 ton of lead ore as collateral
10/1922	10,000	Advance against tax receipts
6/1923	20,000	Advance against tax receipts
12/1923	50,000	Seizure of 5,000 bags by local military
7/1924	50,000	Advance against tax receipts
11/1924	10,000	Advance against tax receipts
1/1925	20,000	Advance against tax receipts
4/1925	30,000	Advance against tax receipts
6/1925	10,000	Advance against tax receipts
9/1925	40,000	Advance against tax receipts

on behalf of Jiuda to end the bidding war was seen as a challenge to Fang's authority and ignored. The company then solicited a letter from Li Yuanhong in the vain hope that the former president's prestige and native place (Li and Fang both hailed from Hubei), or Fan's appeal to another shareholder, Lu Zhonglin (commander-in-chief, Capital garrison command), might help lift the ban.[34] When even such affective ties failed, the company searched for personal access to Fang, finally locating Li Zhaimin, Fang's classmate and son-in-law of a Jiuda branch manager.[35] Other refineries also joined the lobbying effort, including Fengtian, which solicited a cable from Xu Shiying and Tongyi with a letter from Gong Xinzhan.[36] Fang, however, ignored them all, forcing the refineries to cease shipment to the province for nearly a year. During that period, sales of crude salt, however, only increased by 7,000 *dan*, while the Revenue Inspectorate's revenue derived from refined salt shipped to the province declined by over one million yuan.[37]

Jiuda's dependency on network resources also exposed the company to the transitive law of networks: a foe's friend is fair game. With its main production facilities in North China, the company had become privileged in part because of its association with successive governments in Beijing, including the Zhili clique, counting among its shareholders such as Cao Qun and Chen Guangyuan. The two might have their political differences, but a common interest in Jiuda led them to intercede on behalf of the company in several instances. Following Cao's defeat by the Fengtian clique in 1924, however, the tables were turned.[38] Betrayed by the Janus-faced Zhang Hu (see previous chapters), the Hebei Provincial Bureau of Warfare Relief (*Zhili bingzai shanhouju*), with Zhang's son Tongli as associate chief, demanded from Jiuda shares owned by Cao Kun, Cao Rui, Lu Jin, Wang Yuzhi, Wu Yulin, and Liu Menggeng, Gao Ningwei, and Zhang Zhitan on June 15, 1925.[39] Twice the company opened its shareholders registry for examination, but the inspectors could not identify any of the eight war criminals. To facilitate the investigation, the bureau snatched Fan off the street as he walked to work on August 8, 1925. A personal visit by Li Yuanhong, the former president, might have prevented Fan from suffering any bodily harm, but that did not secure his release until he had the company's ledgers and shareholders register delivered to the bureau. He was released the next day, but Zhou Xueting, Jiuda's chief accountant, replaced him as hostage until the company posted a bail of 50,000 yuan.

That only whetted the bureau's appetite. After a thorough examination of the documents, the company was indicted for concealing the identity of *Daxintang* (Cao's household corporation), *Jinxingtang* (Lu Jin's household corporation), and *Xingrangtang* (Wu Yulin's household corporation). As these war criminals had fled with the share certificates, Jiuda must replace them with new ones issued to the bureau. Rules of evidence did not apply: That the war criminals did not use their personal names mattered not (*Daxintang*, the Cao household corporation, was registered under the name of Cao Bingquan). Even the shares of those not named on the list of war criminals such as Li Yanqing were confiscated as well.[40] That both Lu and Wu had sold their shares prior to the bureau's notice to the company was also immaterial. For aiding and abetting the war criminals in hiding their assets, Jiuda was fined an additional 200,000 yuan. The foreign associate chief inspector offered only his sympathy in response to pleas for intervention. Further negotiations settled the charge with another 30,000 yuan.[41]

The refined salt industry, too, was not immune to rent extortion. The saga began in March 1926 when Wu Peifu demanded that the refineries doing business in Hubei match the Huai merchants' 400,000 yuan contribution made in 1925. Chen Diaoyuan in Anhui also followed suit. Revenue farmers in the province had contributed 300,000 yuan to help meet military expenses. The refineries, Chen charged, were profiting from "his" provinces without paying him a share.[42] Wu and Chen threatened to banish the refineries from their turf and seize all inventory. A lobbying campaign led by Kang Youwei (1858–1927) proved futile, as did an appeal to the Revenue Inspectorate in Beijing for permission to apply their payments to the warlords in fulfillment of the gabelle due. Again, the Revenue Inspectorate left the refineries to their own devices, merely reiterating that

> Duties must be paid at this office only. Any payment exacted by whatever officers, civil or military, from the merchants can be by no means considered as salt duty for the central government.[43]

Caught between political disintegration and a principled Revenue Inspectorate, the refineries were forced to down shutters for two months.[44]

Fan Xudong and Jing Benbai worked feverishly behind the scenes for a solution to the problem. They proposed to the Salt Administration, the Revenue Inspectorate, and the provincial authorities a revenue sharing scheme.[45] Fan Xudong appealed to F. W. Mead, the foreign chief inspector, for more flexibility. Sharing the gabelle was still better than nothing at all from the shutdown refineries, he argued.[46] After much wrangling, including a last-minute effort by Jiuda with the chief commissioner Fu Dingyi (1877–1958, also a shareholder of Yongli) in overcoming Mead's reluctance, a deal was struck. Retaining some semblance of its control and functionality, the Revenue Inspectorate continued to collect gabelle from the refineries, but half of revenue was transferred to the provincial authorities upon arrival of the shipments at their destinations.[47] To facilitate the working of the scheme in Hubei, Jiuda drafted a letter for Wei Runsheng introducing itself to Wei's sworn brother, Liu Zulong, the new provincial governor.[48] With the assistance of Fan and his colleagues, the central government's hold on salt revenue achieved by Yuan Shikai and the Revenue Inspectorate was replaced by a de facto revenue-sharing scheme, with Hubei and Anhui receiving half of the gabelle paid by the refineries.

If political chaos forced closures and layoffs for the entire industry, Jiuda bore the brunt in the subsequent round of reshuffling.[49] In North China, Wu Peifu was defeated by an alliance between Zhu Yupu and Zhang Zuolin. Following the precedent of Hubei and Anhui, Zhu, as the new military governor of Hebei, demanded and received a 300,000 yuan monthly subsidy from the Revenue Inspectorate after threatening to seize all revenue from Changlu, virtually the only source of income left to pay for the operation of the Beijing government and the Revenue Inspectorate.[50]

Zhu's success became the envy of his allies. Zhang Xueliang's 3rd and 4th Army Corps occupied much of the salt-producing area of Changlu along the coast, including Tanggu where Jiuda was located. Coveting Zhu's windfall, Zhang imposed a two yuan per *dan* tax on refined salt through the Changlu Salt Consumers' Military Tax Bureau, Peking-Shanhaiguan Section on November 14, 1927.[51] In spite of all salt revenue having been collateralized against the Reorganization Loan, the bureau's jurisdiction remaining local while the company's main market being located in the Yangzi provinces beyond its reach, or whether Jiuda could still compete against other refineries all mattered not, the levy must be paid.

Jiuda was caught again, and the protective cover of its political network largely gone. The Revenue Inspectorate, although not unsympathetic to the company's plight, again refused to allow Jiuda to apply the military tax toward the gabelle as revenue from Changlu had already "dropped off away almost to nothing."[52] Similarly, packing extra salt free of gabelle to compensate for the military tax could not be entertained for fear that the precedent, once set, would open the door to similar demands elsewhere.⁵³ Jiuda bargained with the bureau through Zhou Zuomin and Bian Baimei, and reduced the tax to one yuan and ultimately 0.66 yuan per *dan*.⁵⁴ A 40,000 yuan guaranteed check from Jiuda as a loan to the bureau permitted release of 4,500 tons to the Yangzi provinces, further shipment became impossible when the company and the Revenue Inspectorate failed to agree over whether the loan could be deducted from the gabelle due.⁵⁵ Finally, diplomatic pressure, the inspectorate's threat of withholding the subsidy, and Jiuda's access to Zhu's chief-of-staff arranged through Wang Wendian, chairman of the Beijing Chamber of Commerce, with a gift of 1,000 yuan in Jiuda's company shares, bore fruit. Zhu agreed to share the subsidy with his Fengtian allies.⁵⁶ However, Yan Zepu (1879–1932), finance minister and trusted aide of Zhang Zuolin, was determined to make an example of the company. Juida must pay something, however little. Insisting that it could not compete "with our fellow merchants even if only ten or several cents of military tax (were) to be paid," the company stopped production for over four months until Zhang Xueliang's army finally departed.⁵⁷ Taking advantage of Jiuda's plight, Tongyi and other refineries expanded. Jiuda shipped only 279,080 *dan* in 1927, a 36 percent decline from 1924.

Jiuda's once commanding market share and profits thus began to erode. Pressure from a limited market and predatory competition became so keen that the company's market share in refined salt declined from 95 percent in 1921, when Tongyi began production and broke Jiuda's virtual monopoly, to 26 percent (and 40 percent if we include Yongyu's shipments) by 1927. Average profit per shipped *dan* also slipped from a high of 1.16 yuan in 1920 to 0.819 yuan by 1927 (see Appendix 2).⁵⁸

To the extent that Jiuda owed its success to networking, the strategy was thus not without costs, limitations, risks, and uncertainties. Whether in the form of gifts to Cao Kun and company shares to Li Yanqing or Wang Wendian, they constituted part of the maintenance cost for the company's network.⁵⁹ The search for a tortuous path to its lobbying target, as in the case of Fang Benren, too, suggests that Chinese society of the republican period might be connected by "six degrees of separation" with a sharp cutoff. The company began from its shareholders' network using structural holes to access targets. When that proved ineffective, alternate paths through intermediaries within the intended target's strong network—family, relatives, classmates, and friends—were sought. A letter of introduction, often drafted by the company, facilitated the initial approach. For an imperfect market such as refined salt, the strategy at its best helped alleviate some of the problems of rent extraction, but results could not be guaranteed. Networking also entailed uncertainty and risks, exposing the company to the "dark side" of social capital. As the balance of power between the central government, provincial militarists, and the Revenue Inspectorate shifted, the company led the fight to defend itself and the industry, resetting in the process the relationship between the central and provincial governments, the synarchy of the Revenue Inspectorate, and a sector of the economy in a turbulent period.

Paying for Growth

Success also cost Jiuda in other ways. With a record of profitability, raising capital from the market became an easier task. However, much of the fresh capital did not go to the company. Instead, it financed Fan Xudong's dream of developing China's chemical industry. Founded in 1917 with an initial capital of 300,000 yuan, much of which subscribed by Jiuda shareholders and Jiuda itself, Yongli Soda Ash Company, Ltd. not merely absorbed Fan's attention but also Jiuda's capital, credit, and other resources. Jiuda shared land, dockyard, and senior management with Yongli, with Li Zhuchen alternating with Hou Debang as plant manager while Fan served as general manager without additional renumeration. Beginning with a modest 50,000 yuan in 1920, Jiuda's investment in Yongli soon ballooned as the cost of indigenizing the Solvay process and plant commissioning problems far exceeded Fan's expectation.[60] Beginning in 1922, outstanding bank loan appeared for the first time in Jiuda's annual report despite the 1.2 million yuan in new capital raised the year before. Instead of a net bank deposit of 150,770 yuan in 1921, the company disclosed 87,131 yuan in debt. By 1926, Jiuda owed the banks 680,137 yuan, with Jincheng alone accounting for 567,419 yuan.[61] As one company employee recalled, "much of the additional capital raised by Jiuda actually went to Yongli."[62] By 1933, Jiuda's book value of its investment in Yongli and Yongyu accounted for 45 percent of its paid-up capital (see Table 4.2).

In addition to paid-up capital, available archival sources only offer a glimpse of Jiuda's involvement in the two "younger brothers" in the form of advances diverted from its bank loans or secured by its assets (see Tables 4.3 and 4.4). In Yongli's 1934 massive 5.5 million yuan bond issue underwritten by a syndicate led by the Bank of China, it was disclosed that by 1926, Jiuda had advanced to Yongli

Table 4.2 Jiuda investments (book value) in Yongli and Yongyu, 1921–1937[a]

Year	Yongli	Yongyu	As % of Jiuda Paid-Up Capital
1920	50,000		10
1921	66,000		3.88
1929	120,000	87,500[b]	9.88
1933	592,000	354,500[c]	45.07

[a] Annual balance statements, Jiuda and Yongli.
[b] Yongyu shareholders register, 1929
[c] Yongyu shareholders register, 1933

Table 4.3 Bank loans Jiuda diverted to Yongli and Yongyu, 1922–1937 (yuan, end of year figures)

Year	Jiuda Bank Loans		Yongli	Yongyu
	Fixed	Revolving		
1922	87,131			
1923	411,210			
1924	278,829			178,000
1925	459,147			228,044
1926	680,137			227,000
1927	586,592			218,286
1928	645,768			218,756
1929	250,000	795,417	464,974	210,470
1930	250,000	743,197	585,403	202,526
1931	250,000	733,513	503,065	162,000
1932	160,000	669,945	450,000	162,000
1933	200,000	551,498	360,000	67,500
1934	150,000			67,500
1935	80,000	448,064		35,200
1936		123,871		35,200
1937		345,926		

over 1.3 million yuan without interest. Only in 1929 did the company began charging an annual interest of 5 percent (half the market rate).[63] In addition, the "younger brother" also took loans (see Table 4.4) directly from Jincheng Bank secured by all its assets and a secondary guarantee by Jiuda.

Jiuda's shareholders began to complain as interest charges eroded profits, forcing the company to dip into reserves to maintain an 8 percent dividend rate (see Appendix 2). Coming under heavy criticism in

Table 4.4 Loan from Jincheng Bank to Yongli (with secondary guarantee by Jiuda)

Year	
1922	130,022
1923	
1924	149,854
1925	166,270
1926	604,784
1927	
1928	
1929	480,000
1930	77,811
1931	50,372
1932	121,907
1933	61,601

an annual meeting, an undaunted Fan Xudong pleaded with dissatisfied shareholders: in deploying the capital entrusted to him, he must be socially responsible, including promotion of the country's chemical industry. Whether Jiuda was declaring dividends of 8 or 20 percent, he would have made the same decisions.[64] Fan survived the vote and continued as general manager and board member of the company.

Recalling those turbulent years, Fan Xudong succinctly summarized Jiuda's problems in 1928. Such was the nature of the salt business that he found himself preoccupied with lawsuits and dealing with politicians and bureaucrats.[65] He and his colleagues addressed organization and management problems as best as they could through a hybrid strategy combining market, hierarchy, and network. But with Tongyi equally well-heeled and the Yingkou refineries enjoying lower production costs, Jiuda struggled even as refined salt came to command over one-third of the supply in Hankou and Jiujiang by 1929.[66] Revenue farmers, too, defended their privileges vigorously, their legitimacy and influence renewed through underwriting loans, advances, and contributions. Officials at the Salt Administration and the Finance Ministry, as corrupted or conscientious public servants protecting an important source of state income, took the revenue farmers' charges seriously by enforcing the law limiting refined salt to the commercial ports, however defined.[67]

Securing the salt fields of Qingdao through Yongyu presented both opportunities and challenges. Now that the group controlled over half of the country's production capacity in refined salt, integrating

Yongyu's production and sales also proved daunting. In addition to satisfying two overlapping but not identical sets of shareholders, combining Jiuda and Yongyu for economy of scale also raised agency problems with local franchisees. Construction and plant commission problems at the Yongli, too, preoccupied Fan Xudong and diverted much of Jiuda's resources to the new enterprise, exhausted its credit line, and strained its relations with Jincheng.[68] To a determined Fan Xudong, however, that was a small price to pay for realizing his dream of developing the country's chemical industry.

5
Cartel as a Business Network

As Jiuda adapted its campaign of creative destruction against the revenue farming system, overcapacity and chaotic, if not wasteful, bargaining price wars among the refineries led to adoption of further restrictive strategies. Not unlike other entrepreneurs in China then and elsewhere confronting similar issues, the refineries resorted to collusion for collective survival.[1] Beginning with a local ad hoc collaboration, the salt refineries' efforts culminated in the establishment of the National Salt Refineries Board (*Jingyan zonghui*, hereafter NSRB) in 1926. Using agreements to limit production, pool sales, coordinate pricing, and with elaborate compliance rules, as well as lobbying with the government to impose barriers of entry against new refineries, they learned how to manage the market through cartelization from 1926 to 1934.

Was this interfirm network successful or unsuccessful, or both? Economics textbooks have little good to say about cartels; collusive agreements (written or otherwise) among firms to raise prices, determine output, allocate and divide markets, and restrict new entry violate the principle of market competition and result in less than optimum consumer welfare. Any such efforts are also doomed. Asymmetry among participating firms, and problems of self-interest, cheating, coordination, free-riders, enforcement, and punishment make such attempts unsustainable in the long run.

But then, how "long" is long? Some cartels lasted as much as a century, nor should periodic bargaining price war be interpreted as a sign of failure.[2] If most cartels manipulate the market for inflated profit, price data and other empirical evidence remain open to interpretations as economists, accountants, and lawyers debate how to measure cost (marginal, average, incremental, or total) and hence profit. In the event of market failure, should collusion be permitted at the cost of reduced competition?[3] The master narratives of free market competition and trust-busting must be reconciled with economic history.[4]

The same is true in the case of modern China, an economy characterized by a "generally unrestricted competition" notwithstanding. There is evidence of collusive activities and cartels, whether informally or formally organized, local or national, domestic or foreign, imports and exports. From cotton cloth trading in Dingxian to cement, matches, flour, egg products, coal brisket, enamelware, and soda ash, some of these attempts are more elaborate than others, although little is known about their organization and operation.[5] Leaving the role of the state and the issue of the refineries' success or failure (or both) in lobbying the state to the next chapter, this chapter studies the behavior of the refined salt cartel. As a network, the salt refineries shared various attributes and a common interest against revenue farmers and rent-seeking officials. However, with crude salt as a close substitute (albeit allegedly an inferior one), competition among themselves and local distributors, the refineries' profit margin remained slim. Amid political disintegration, a declining Revenue Inspectorate did little to protect the industry from rent extractions, even though the refined salt industry might be its best customer. Instead of a standard textbook example of consumer-gouging cartel, the refined salt industry increasingly sought collective survival by solving, to the extent possible, problems of bargaining, compliance, and network multiplexity as its ranks grew.[6] To deal with all these complex problems, Jiuda continued to adapt by evolving a centralized business organization combined with networks.

From Competitors to Collaborators

Organizing the refined salt industry began locally in 1925. Shipments from new refineries in Yingkou were set to arrive in Hankou. At the same time, authorities in Shandong were demanding contributions and stamp duty on Tongyi's shipments.[7] Both Jiuda and Tongyi thus felt the need for a truce.[8] In September, Xu Zhaozhou, Jiuda's Hankou manager, began discussing a limited liability partnership with Huang Wenzhi, Tongyi's local franchisee.[9] It was a marriage of convenience. Huang, president of the Wuyue Chamber of Commerce and managing partner of Tongyi's Hankou franchise, wanted new partners, especially with Yongyu and other refineries poised to enter the Hankou market. Matching Tongyi's marketing strategy had proved costly and ineffective for Jiuda. Joining forces might reduce that cost and bring in the resourceful Huang while strengthening the companies' ability to deal with unruly local subfranchisees and the state. Both companies agreed to keep a low profile, hoping that Datong, the new partnership, registered with only 100,000 yuan in capital might present a less inviting target for the provincial authorities.[10] Hengfengtang, Jiuda's holding corporation, invested 55,000 yuan and Tongqing (Tongyi's subsidiary) contributed 45,000 yuan to take over as the refineries' franchisee. The partnership contracted for an annual quota of 150,000 *dan* of refined salt from Jiuda and Tongyi, as well as 80,000 *dan* from Yongyu. Huang, in exchange for agreeing to Jiuda's larger market share, served as the general manager.[11]

Realignment of provincial politics following the death of Xiao Yaonan prompted the refineries to seek collective security by organizing themselves. Chen Jiamo, the new Hubei military governor, demanded 400,000 yuan from the refineries to match the contribution made by Lianghuai revenue farmers.[12] Confronted by Chen's threat of a ban, Datong, Liyuan, and Fengtian formed a refined salt association (*Jingyan gonghui*) to arrange an advance to the provincial treasury from Datong Bank (unrelated to Jiuda and Tongyi's Datong) against future-dated provincial gabelle receipts.[13] This arrangement permitted the refineries, as their shipments arrived, to redeem the receipts from the bank as evidence of gabelle payment. Spared individual responsibility for the loan, the refineries collectively were also assured of continued access to the market with the quid pro quo—Chen could not renege on the deal and ban refined salt with the gabelle already advanced.

Realizing that their survival, individually and as an industry, was threatened, the five refineries decided to organize themselves formally. They petitioned Beijing for the establishment of the NSRB in April 1926, pointing to the industry's important service to the state: it contributed over a tenth of the gabelle each year. Unlike revenue farmers, however, the refineries were scattered around the country, subject to different jurisdictions with varying weight standards and prices. Consulting each refinery on official matters was logistically difficult and time-consuming. The petitioners proposed the formation of a national association headquartered in Beijing to promote collaboration while facilitating communications with the state. In their public manifesto, they also declared war against the revenue farmers. They contended that for the national good, the consumption of adulterated and unhygienic salt must be stopped and the monopoly system dismantled.[14]

The revenue farmers reacted with fury. Launching a national media and letter-writing campaign, they rebuked the upstarts for jeopardizing the livelihood of thousands of salt producers. In the name of science and technology, the refineries were colluding with warlords and corrupt officials to squeeze revenue farmers, producers, and consumers for their own enrichment.[15] Other supporters chimed in, observing that the refineries were only trying to recoup their losses.[16]

The refineries survived the counterattack. From Beijing's perspective, the NSRB was no different from the Huai revenue farmers' Yangzi provinces guild in performing services deemed necessary to the state. Registered as a trade association (*tongye zonghui*), the new organization, with a twenty-two clause charter, was charged with the responsibility of presenting the industry's opinions on the formulation, revision, and review of the salt code. It was also empowered to advise the state, mediate (but not adjudicate) disputes among members, conduct and publish surveys, and pursue any other matters beneficial to the industry.[17]

Economics of Collusion

One benefit took the form of an elaborate pooled sales scheme. Fengtian and Liyuan joined Datong to market their salt through the local chapter of the national association as an experiment for one year beginning June 27, 1926.[18] Datong's share of the market totaled 430,000 *dan*, leaving 100,000 *dan* each for Fengtian and Liyuan.[19] Each refinery (or its franchisee) took turn to sell at a price set by the chapter. Representatives of the refineries, as unpaid officers of the chapter, monitored the arrival, storage, and disposal of shipments to ensure compliance.

However, tension and conflict marred the cartel from the beginning. Scattered around the country, different jurisdictions shaped the refineries' behavior. When the Nationalists on their Northern Expedition reached eastern China, they disrupted crude salt supply from the coast. Zhang Qinji, Fan's brother-in-law (and Jiuda shareholder), spotted the resulting shortage in Hunan and cabled Fan on October 24, 1926, with an urgent initial order for 40,000 *dan* of crude salt from Yongyu. It was a risky proposal—the shipment needed approval by the Salt Administration and the Revenue Inspectorate in Beijing for a suspension of the divisional boundary, a revenue-sharing arrangement negotiated between the central government and the provincial authorities, not to mention shipping through a war zone.

Without sharing the risk (or the potential profits) with other cartel members, Jiuda exploited the opportunity alone. After tortuous bargaining, Beijing approved shipments totaling 300,000 *dan* with a regular gabelle of 4.5 yuan per *dan*, special fees of 1.2 yuan per *dan* paid to Pan Fu (then prime minister), 0.30 per *dan* to Fu Dingyi, and 0.25 yuan per *dan* to the Hunan Provincial Salt Bureau for its cooperation.[20] Xinfu, a new company with no formal ties to Jiuda or Yongyu, handled the entire transaction with shipments in foreign bottoms. The arrangement reportedly netted Fan over 2 million yuan, although expenses were also considerable.[21]

Whatever Fan gained from this transaction also cost him. The use of an alias, coded communications, and Xinfu did not fool his fellow cartel members. Tongyi's local franchisee attempted to follow suit with 90,000 *dan* shipped from Yantai without making the necessary arrangements, resulting in confiscation of the shipment. Whether out of jealousy or competition, or both, Jiuda and Yongyu became the common target of their fellow refineries.

Continued cooperation became strained. In February 1927, citing rising costs of coal and doubling of the price of crude salt and gunny bags, Jiuda notified Datong of its decision to raise its price by 0.5 yuan per *dan*. Replaying the agency problems that Jiuda had experienced with its franchisees elsewhere, Huang Wenzhi rejected the increase. With refined salt already more expensive than crude salt, the franchise could not possibly absorb the raise.[22] Jiuda threatened to stop delivery, reminding Huang that the two would perish together. In reply, Datong acknowledged that while production costs had been rising in the North, Hankou had its own share of problems. The provincial Salt Bureau had demanded a loan of 1 million yuan from the industry, followed by another order to purchase 200,000 yuan in bonds.[23] Although the partnership remained profitable, with Jiuda's share of the profits for 1927 amounting to 165,180 yuan, both sides understood that their partnership was over by February 1928.[24]

Jiuda formed a new subsidiary, Taiyu, serving several purposes for the all-important Hankou market. Registered as a separate entity, Fan and his colleagues again hoped that this might present a less tempting target for rent seekers. It might also minimize marketing cost by combining Jiuda and Yongyu wholesale operations.[25] Tanggu also improved its quality control, with each bag of salt weighing over 217 lbs., net of tare and tret, a windfall of at least seven lbs. *gratis* to attract customers.[26] Shipment of refined salt to Hankou surpassed that of crude salt for the first time in 1929, although price-cutting and various levies also combined to reduce Jiuda's average profit to 0.43 per shipped *dan*, or a 63 percent decline from its peak a decade ago (see Appendix 2). With warfare between the Guangxi faction and Jiang Jieshi disrupting the Hankou market, a workers' strike at Tanggu against both Jiuda and Yongli demanding pay raises, over half a million yuan in bank loans coming due and few receivables, the company had had to draw down its reserves to maintain an 8 percent dividend for the year. Business was bad.[27]

Resumption of Cartel, 1930

So, too, was competition. The refineries learned the lesson that price-cutting and packing extra salt for free could not continue, especially with a new rent-extracting regime. In 1929, the Hubei Provincial Salt Bureau again demanded another 1 million yuan advance.[28] With the heavy losses (or reduced profits) sustained in the previous two years, the refineries collaborated again. In February 1930, the general manager of Tongyi and representatives of the six refineries at Yingkou proposed to travel to Tianjin for an exploratory meeting.[29] Writing to Fan Xudong, Fan Gaoping argued that critical issues threatened the industry's survival. With shifts on the national political stage and pressure from revenue farmers, continued rivalry among the refineries and their local franchisees spelt doom for the industry—the government's grabbing hand (*juling zhizhang*) would seize every cent.[30]

Jing Benbai supported the initiative enthusiastically, envisioning a national cartel of salt refineries led by Jiuda and Yongyu. Inspired by the domination of big powers in world politics, he proposed a grand alliance with Tongyi and Liyuan to form a national trust, forcing the smaller refineries into line or driving them out of business.[31] Collaboration of the refineries also addressed an agency problem created by some local franchisees playing the refineries against each other. Jing proposed the formation of sales cooperatives to replace local franchisees which had been forcing price concessions from the refineries even as they enjoyed lucrative commissions. Henceforth, all the participating refineries shared the expenses and profits according to their sales. Following the business practices of foreign companies in China, prices were adjusted by telegraph and local cooperatives and retailers subsidized by the cartel to ensure their loyalty.[32]

Fan, too, had other considerations. Jiuda was under pressure on several fronts. Over 1.3 million yuan of its available capital and line of credit had been diverted to finance Yongli in a battle with Imperial

Chemicals Industries, Ltd. for the soda ash market in China and Japan.[33] As Yan Xishan gained control of Hebei, he also attempted in February 1930 to impose a Changlu consumption levy of one yuan per *dan*. Jiuda again ceased operation for two months until Yao Yongbai, an early company shareholder, intervened in his capacity as Yan's financial adviser.[34] Surviving on loans from various banks and overdrawn credit lines, Jiuda could ill afford a war on another front.[35]

When they finally met in Shanghai, representatives of the major salt refineries still found much disagreement. While acknowledging that the cartel saved the industry in 1926 and that there was a need for a united front against local franchisees as well as revenue farmers and the state, they disagreed on how to solve the problems of overcapacity in a restricted market. Fan Gaoping, representing Liyuan and other Yingkou refineries, wanted the government to open new markets to absorb the excess capacity. Fan Xudong took a different tack and argued that the refineries should collaborate to limit production and lobby the government to impose a moratorium on new refineries.

The refineries also disagreed over the institutional framework of the proposed cartel. In addition to Jing's local cooperative scheme, they also considered the formation of a National Refined Salt Co. (*Zhongguo jingyen lianhe gongsi*), with a capitalization of 5 million yuan to pool market all refined salt in the country. When this proved overly ambitious, Fan Xudong proposed consolidation of wholesale operations in various markets, a uniform price of 2.6 yuan per *dan* (factory F.O.B.), and a minimum profit of 0.6 yuan per *dan*.[36] Perhaps Fan hoped that this arrangement would inspire cooperation.

It did. After a series of contentious meetings that lasted over a month, the refineries forged an elaborate sales pact (*yunxiao gongyue*) with different prices and ratios for each company in various markets for one year (see Appendix 7). At Fan's urging, the refineries also addressed the problem of compliance and resolved that any violation be subject to confiscation of the offender's contribution to the association's endowment. With the promise of collective security, sharing the costs of opening new markets, and a stable share of the market and profits, the refineries fell into line. The Joint Office of Salt Refineries (*Jingyan gongchang lianhe banshichu*), headquartered in Shanghai, was established to coordinate and enforce the pact with Fan Xudong elected to chair its executive committee. Pooled sales in Wuhu and Hunan began on May 4, 1930, followed by Jiujiang on May 11, 1930, and Hankou on June 1, 1930.[37]

Cartel in Action

The new cartel, learning the need for hierarchical control and coordination from its past experience, performed a variety of duties: enforcement of the pact, pooled sales, removal of uncooperative local franchisees, price fixing, and lobbying officials on matters of collective concern.

As its first task, the association dispatched representatives to assist various chapters in adapting its policies to local needs. In Hankou, the local chapter took custody of inventory and sales as a first step. By August, the refineries finally adopted a detailed arrangement for pooled sales and profits. Each took turn to sell at a price set by the chapter to customers who must settle by bank orders with a grace period of no more than two weeks. Every ten days, delegates from the refineries took inventory, examined the bills of lading, and verified deposits, calculated at one yuan per bag sold, to the profit pool account in designated banks. At the end of each month, the deposits were distributed among the refineries according to their market share, hence reducing the problem of cheating. In addition, if delivery fell below half of a refinery's quota, only half of its deposits was refunded, with the remainder going toward the national association's endowment.[38]

Asserting their renewed unity, the refineries began pooled sales to eliminate local franchisees and to raise prices. Taking advantage of the chapter's marketing power in Hankou, Fengtian did not wait for its contract with an uncooperative franchisee to expire. The company notified the Joint Office of its intention to stop delivery to the franchisee and, with the local chapter handling the sales of its shipment, requested that all proceeds be withheld on its behalf. At the same session, the Joint Office also discussed a proposal from the Yingkou refineries to increase the price by 0.2 yuan per *dan* to compensate for rising cost. Approved unanimously, all the local chapters were notified by cable, with the increase going into effect November 1, 1930.[39]

Not all the local chapters, however, heeded the Joint Office's decisions. The franchisee holders of Anqing, together with their network of distributors, defied Shanghai's order. Fan Xudong traced the problem to Tongyi and its local franchisee. Although Jiuda pioneered the access to this market, Tongyi and its franchisee had become the biggest supplier (25 percent in 1930 and 52 percent in 1931, as opposed to Jiuda's 15 and 26 percent, respectively) on the strength of their overpacking and generous credit policy. Instead, they demanded that Jiuda's share of the market, as negotiated in Shanghai, be reduced by eight, and Yongyu assigned only six shares because of its poor quality. To prevent its market share from further erosion, Jiuda agreed to a secret price reduction of 0.2 yuan per bag (or a Tanggu F.O.B. price of 2.1 yuan) as requested by Bao Yanbin, its local franchisee. Much to the displeasure of its Shandong shareholders, Yongyu also had had to accept a lower price and absorb the loss despite continued improvement of its product.[40] Issues such as sodium content, color, and packing weight also made pooled sales difficult. At the behest of the Joint Office, Fan Gaoping and Fan Xudong again visited several chapters to help convince local distributors of the need for standardization.[41]

Despite all these problems, Fan Xudong considered the cartel's first year of operation smooth. Jiuda's net profit per shipped *dan* improved from 0.43 (1929) to 0.618 yuan (1930) despite a low volume of 229,320 *dan*.[42] Again, cartelization helped the company survive a difficult time.

Renegotiations of 1931

This modest improvement, however, was achieved at considerable cost to Jiuda. For the first time, Tongyi led the industry with sales of 261,624 *dan*. Although Jiuda reclaimed that title in 1931 with a total of 263,424 *dan* shipped, edging out Tongyi by a slim 1,890 *dan*, it still trailed Tongyi in the Hankou and Jiujiang markets while its net profit per shipped *dan* declined to 0.57 yuan (see Appendix 2). Jing Benbai was aghast at Jiuda's performance and idle capacity. Plagued by problems of coordination, enforcement, and the games that refineries played against each other, it was time to renegotiate.

In a Joint Office meeting on May 14, 1931, attended by representatives of all the refineries (but not franchisees), Fan Xudong challenged the network: how could pooled sales continue? Lin Zizhen, assistant manager of Tongyi, emphasized the need for honesty in the collaboration. The industry had benefited from pooled sales, but to demonstrate unity and commitment, the refineries needed more stringent regulations against any violation. Yin Zibai, Wuhe's manager, argued that operations of the Joint Office must be improved to avoid the harmful effects of competition. Xia Zhubo, manager of Fuhai, echoing Fan Xudong's position in his negotiations with Tongyi in 1930, urged his colleagues to focus on enlarging the market rather than fighting over market share.

Xia's plea fell on deaf ears, however. While the refineries were unanimous about continuing the cartel, they continued to clash over market allocation. Yuhua insisted on a larger share in Jiujiang as a reward for its local franchisee's effort in developing the market. Dismissing the implied criticism that they were free-riders, Fan Gaoping countered that joint sales benefited everyone with stable prices and profits. Lin Zizhen concurred and argued that even cornering the market might prove unprofitable because of price cutting. With Tongyi in mind, Fan Xudong observed that someone would have to make concessions to satisfy the demands of the smaller refineries; Jiuda could not be expected to reduce its market share alone. After hard bargaining, the major refineries voted to keep the existing allocation ratio for Hankou and Jiujiang while Yuhua and Tongda held out for better terms. To placate them, the major refineries offered concessions in Hunan. After more bargaining, the four refineries (Jiuda, Tongyi, Liyuan, and Wuhe) conceded 35 percent of the Hunan market to give each of the six smaller refineries seven shares.[43]

In contrast, organizing the Anqing chapter proved to be the most daunting. Tongyi and its franchisee, on the strength of their generous credit policy and overpacking, had captured over half of that market, and other refineries were reluctant to concede this rapidly growing market. Fan Xudong made clear that Jiuda had pioneered access to this market at considerable expense and should thus be rewarded with a larger share. However, Tongyi rejected any concession. Taking the high road, Fan Gaoping offered to give up Liyuan's share (less than 2 percent) to complete the cartel. While this passive-aggressive tactic compelled all the refineries to reaffirm market sharing in principle—which meant that Liyuan must not give up its allocation—no one budged either. Deadlocked, the refineries agreed to divide the Anqing market into 110 shares instead of 100, effective July 1, 1931.[44]

With the fight over market shares resolved for the time being, the refineries turned their attention to compliance issues. A new set of regulations specified stringent penalties against secret price cuts, extended grace periods, overpacking, and selling beyond one's allocation. Retaliation would be swift: a price reduction collectively or a subsidy to those which did to enforce order in joint sales.[45]

Problems continued to plague the refined salt cartel, however. Taking advantage of the protracted negotiations and a loophole in the new regulations—there was no provision for the amount of inventory already ordered by distributors or in transit—Tongyi's local franchisee flooded the Anqing market with a 1,700 ton shipment in June. A brief war erupted as the other refineries retaliated, lowering their price by fifty cents below that charged by Tongyi. Following another round of bargaining in Shanghai and intervention by the Anqing Chamber of Commerce, the refineries finally signed a new pooled sales agreement on August 6, 1931. Further shipment to Anqing was suspended until existing inventory was cleared according to the sharing ratio. While order was restored, the resolution did not mention any penalty on Tongyi or its franchisee.[46]

Bickering and compliance issues thus marred Jiuda's cartel experience. Did the refined salt cartel fail in the short run then? Short-lived price wars, as part of the bargaining process, did not mean the end of the cartel. Shipment record for 1930–1932 (percent of sales by refinery) for Hankou and Jiujiang, the two main markets, correlates highly with the negotiated market share (+0.91 and +0.9, respectively). Statistically, the refineries thus abided more or less by their agreement. However, in both markets Jiuda's average sales were below its allocation (Hankou: 15.7 percent of the market instead of 18.3; Jiujiang: 15 instead of 15.18). To placate the smaller refineries, Jiuda actually conceded market shares.[47] Tongyi, in contrast, shipped more than it was entitled to (Hankou: 19.3/15.4; Jiujiang: 22.9/17). Contrary to Jing Benbai's expectations, Jiuda was vulnerable because of its circumstances and commitments. If the success of the cartel depended, in part, on concessions from Jiuda and Tongyi, Jiuda carried more than its share of the burden. It also suggests that the cartel made it easier for the smaller refineries to survive.

Sons of Hunan

As the refineries continued their games, holding out for more market shares and raising prices where joint sales remained in effect, Fan Xudong's frustration mounted.[48] Taking advantage of the accommodation made for the smaller refineries, Hongyuan split itself into two companies in an attempt to claim more market share.[49] Bickering over market allocations aside, with pooled sales came collective responsibility for the gabelle and any other levies due, a risk that Fan Xudong and his colleagues found unacceptable.[50] He also opposed another Tongyi proposal to raise the price by 0.2 yuan per *dan*. Manufacturing costs undoubtedly had been rising, but Fan argued in vain that prices must be kept low against crude salt.

Long-term survival of the industry now preoccupied Fan Xudong. Miao Qiujie, since his appointment on October 18, 1930 as Lianghuai commissioner, had imposed a series of reforms to raise more revenue from the division. Starting in Hankou, Miao's "restricted free trade" policy established a state-set monthly sales quota and price ceiling on crude salt. Upon delivery and payment of all gabelle and levies at designated banks, revenue farmers no longer needed to wait in turn but were free to sell under the price ceiling. A faster turnaround of capital, generous bank financing of gabelle payment and document billing, construction of new state depots, antismuggling measures, and the shift to steamship transportation improved the crude salt business. Serving concurrently as chief of the revenue guards and held accountable for the delivery of Lianghuai revenue, much of which was generated by crude salt, Miao also enforced regulations and restrictions on refined salt with renewed vigilance.[51]

Miao's policies thus posed a serious challenge to Fan and the refined salt industry, the two men's personal rapport notwithstanding. Attempts to expand the market beyond the commercial ports were repeatedly denied, revenue guards increased their effort, and, as crude salt prices declined with improvement in quality, refined salt increasingly became a luxury. The market for refined salt, especially in Hubei, began to shrink, as shown by Table 5.1.

Any price increase was thus suicidal. On the other hand, lowering the price, whether as a result of competition among the refineries or to reclaim market share, also was not feasible. Given the different cost structure, the revenue farmers enjoyed a much higher profit margin (see Table 5.2) and far deeper pockets. If competition among the refineries drove the price down too far, the revenue farmers could (and did) complain that refined salt was encroaching on the crude salt market and disrupting the stability of state revenue.

Table 5.1 Crude salt versus refined salt shipment: Four Yangzi provinces, 1931–1936[a] (Unit=Percentages; C=crude, R=refined)

Year	Hubei		Hunan		Jiangxi		Anhui	
	C	R	C	R	C	R	C	R
1931	59	41						
1932	66	34	92	8	67	33	87	13
1933	79	21	90	10	52	48	89	11
1934	84	16	90	10	72	28	94	6
1935	81	19	85	15	67	33	78	22
1936	85	15	90	10	64	36	82	18

[a] Tabulated from ZGYZSL, 2nd comp. (1943), I: Xiang'an, 5; E'an, 21–22; Xi'an, 1; Yuan'an, 6.

Table 5.2 Cost comparison of refined salt and crude salt, Hankou 1933[a]

Refined salt	
Cost: Production tax due in Tianjin	1.5
Transit tax due in Shanghai	1.5
Division tax due in Hankou	1.5
Additional levies due at Hankou	6
Foreign loan in gold supp.	0.3
Depot construction fee	0.1
Factory price	3.1
Shipping	0.7
Joint Office fee	0.13
Gross cost per *shi*	14.8
Income: per bale of 150 catties Less commission 0.2/bale	23.6
Or	15.73 per *shi*
Gross profit for refined salt	0.93 per *shi*
	12,000
Huai crude salt	
Cost: Gabelle per ticket of 4,000 *shi*	
Additional levies	32,550
Foreign loan in gold supp.	1,302
Depot construction	434
Production/shipping	7,400
Ticket rent (if applicable)	500
Huai Guild charge	800
Misc. depot fees	1,200
Subtotal	56,186 or 1.85/*shi*
Income: including gabelle-free extra-weight	63,715.9
Gross profit: not including local expenses	7,529.9 or 1.88/*shi*

[a] *Source:* JD 100954, 100981, 100982.

Disagreement over pricing policy aside, Fan was also confronting a rapidly changing refined salt industry. Japan invaded the Northeastern provinces in September 1931, and with that, the six refineries at Yingkou fell under Japanese control together with the divisional salt administration. Any gabelle revenue generated there, Fan reminded the Salt Administration, would be financing the invaders.[52]

With the country and the company's fate intertwined, Fan went on the offensive. Yongli was finally turning the corner with a price-fixing agreement with Imperial Chemical Industries and a 2 million yuan debenture syndicated by Jincheng and other banks. Although much of the issue remained unsold, Jiuda took 500,000 yuan of the paper which banks accepted as collateral for loans. With the pressure on Jiuda's finances easing somewhat, Fan attacked.

When the refineries began discussing renewal of the cartel in July 1932, Fan Xudong made his demands clear. He had promoted the cartel to prevent unnecessary competition (*wuwei jingzheng*). His company, with almost half of its production capacity idling, had sacrificed much in the hope that the industry might work together to expand the market. Instead, the refineries wasted their energy fighting over market allocation as entitlements. Henceforth, Jiuda could no longer afford the sacrifice; he demanded an annual sales of at least 300,000 *dan* (Hankou: 120,000; Jiujiang 100,000; and Anqing 40,000), or a quarter of the refined salt market, to maintain an 8 percent dividend rate. Specifically, this translated into 5½ additional shares in Hankou, eight more shares in Jiujiang, and two more shares in Hunan.[53]

Fan's blunt demands caused a scramble among the refineries for a solution.[54] Tongyi was determined to protect its gains and remained noncommittal about the renewal, if only to force concessions from the smaller refineries. The Yingkou refineries, led by Wang Xuhe of Fengtian, wanted to continue the cartel and preserve their share of the market. After prolonged discussions, Fengtian and Liyuan agreed to concede one share each, and Huafeng and Minsheng half a share each in all the four Yangzi provinces to Jiuda, with Fuhai, Yuhua, and Hongyuan each compensating Minsheng with a side payment of 500 yuan. Four days later, Fan Xudong declined the offer. While he appreciated the goodwill, he did not want any concessions from Fengtian, Liyuan, and Huafeng. Instead, he urged Wang to find a more equitable solution.[55]

Two days later, Wang reported the result of his negotiations with Lin Ziyou: Tongyi would not insist on an equal share of the market with Jiuda. Instead of Fengtian and Liyuan conceding their market shares, Jiuda and Tongyi's allocations would be increased by three and one, respectively. Wang conceded that the offer might not satisfy Jiuda, but it would preserve the industry's integrity and the possibility of working together to expand the market at a time of national crisis. It was the best he could do.[56] A disappointed Fan decided to force the issue, expecting no more significant concessions. He cabled Wang three days later rejecting the offer. Beginning August 1, 1932, Jiuda would withdraw from the cartel.[57] The threat worked as the Yingkou refineries, already under pressure because of the Japanese invasion, offered concessions to preserve what remained of their market share through the cartel.[58]

Through the bargaining, Jiuda gained not merely market share but also an opportunity to reorganize itself. The price wars exposed the company's weaknesses. Pooled sales had rendered its sales staff complacent, and internal memos pointed to a lack of control over factory cost and inventory.[59] Jiuda confronted the problems with another round of internal reorganization, again combining hierarchy and network. As his group grew, employee factions, based variously upon native place or schools, had emerged. The Hunan faction led by Yu Xiaoqiu (same report identified Yu as Fan's proxy), a graduate of the University of Chicago, dominated senior management. In turn, Jing Benbai, Chen Diaofu (1888–1961, a chemist trained at Suzhou University, chief engineer of Yongli, 1917–1922), Hou Debang (1890–1974, a chemist trained at MIT and Columbia, who succeeded Chen), and Tong Gengran (1902–1954, Yongli plant manager, 1948–1952) each had their supporters among the staff and workers from Zhejiang, Jiangsu, Fujian, Hebei and Shandong, respectively. Graduates of Nankai University and Hebei Higher Technical College, too, had their allegiances. Efforts to establish a personnel department thus encountered resistance, and for many years Li Zhuchen, a chemist trained in Japan and another Hunanese, was in charge of personnel matters, Jiuda's production, and rotated with Hou Debang as Yongli plant manager to maintain a delicate balance.[60]

Fan and his colleagues thus became keenly aware of the need to exert control through formal governance (*fa zhi*) instead of personal ad hoc decisions (*ren zhi*) to overcome factionalism, strengthen the company, and forge a group identity.[61] Elaborate rules were crafted and issued to employees to ensure smooth coordination among various units through a joint office established on June 1, 1932, with a 100-clause manual for Jiuda's workers. The group became nationally known for pioneering an eight-hour work day, free medical care, maternity care, and education for employees' children, subsidized housing, utilities, and mess hall, not to mention a generous savings scheme and bonus system.[62]

Such measures became an integral part of Jiuda's corporate culture, commitments, and representation as a patriotic enterprise. In 1934, Fan Xudong invited his "family" to formalize a set of guiding principles, if not a constitution, for the group. Among his fourteen suggestions were a belief in mutual help, no obstacle that could not be surmounted, the revival of China in twenty years, and the insignificance of personal gain. The editors of *Haiwang*, the group's journal, summarized the many contributions into four principles: an absolute dedication to science, promotion of industrial development, the need of the group outweighing that of the individual, and serving society as a privilege.⁶³ Some might criticize these principles and the additional expenses as part of the capitalist's scheme to manipulate and exploit labor, but Fan spared no expense on them.⁶⁴ As an investment in corporate identity and loyalty, it was a small price for the apparent success in addressing factionalism and agency issues confronting the group. He also recruited Li Tifu to head Jiuda's marketing office. A fellow Hunanese and a graduate of the University of Illinois, Li reorganized the company's sprawling operations, including the introduction of a Western accounting and auditing system for tighter cost control. Through these reforms, Fan challenged his plant managers to meet a target factory cost of 1.6 yuan per *dan*.⁶⁵ It cut costs by moving the downsized Shanghai and Hankou regional offices to cheaper quarters. To energize its sales staff, the company also switched to a commission system.⁶⁶ Adaptation and institutional innovation was still possible even for a company in a cartelized industry.

As reconstructed above, the story of the NSRB offers a rare glimpse into the operation of a cartel in modern China. It exhibited many features of collusion: price fixing, pooled sales, and market sharing as modeled by economists. Not unlike its counterparts in the West, self-interest, cheating, loopholes, short bargaining wars, and compliance issues disrupted the cartel as it learned to grapple with the problems. Penalty clauses, increasingly elaborate and stringent, were rarely enforced as the NSRB did not have the legal authority to adjudicate. However, in Jiuda's case, collusive action did not yield supranormal profits. Similar to the experience of other cases in Europe then, the company, despite its size, found it necessary to concede market share to the smaller refineries to prevent uncontrolled collapse of the industry, wasted capital, and unemployment.⁶⁷ Tongyi, the second largest refinery, played the game well, sitting on the fence by playing Jiuda against the smaller refineries while enjoying a dividend rate of 20 percent from 1932 to 1934.⁶⁸ For all its faults, cartelization, as a "higher" form of economic organization that "replaced the brutal ethos of competition with a system of cooperation," helped an industry to survive in hard times.⁶⁹

On the other hand, the NSRB experience also highlights the social embeddedness of the salt refineries. Whatever their disagreements as they learned to manage competition through cartelization, their bargaining and negotiations were shaped by common attributes such as geography (such as those in Yingkou), or native place and work experience (Fan Gaoping, for instance, had worked for Jiuda as manager of its Nanjing office in addition to his work with Jing Benbai at the

Zhejiang provincial salt bureau in 1911–12). As an entrepreneur, Fan Xudong managed not merely Jiuda as a hierarchical business organization, but also the overlapping networks to realize his dream of national salvation through refined salt and development of the chemical industry. Tunway became a shareholder of Jiuda, and Yu Xiaoqiu, Fan Xudong's chief aide, held fifty shares of Tongyi; Fan Gaoping, Huang Wenzhi, Yin Zibai, Xia Zhubo, and Zhong Ziliang (chairman of Dinghe) all invested in Yongli, owning 15, 73, 15, 29, and 15 shares, respectively.[70] The refineries thus continued their game while also working together to further their common interests by lobbying the Nationalist government, as we shall see in the next chapter.

6
Relationship with the Nationalist State

The relationship between the Nationalist government and private capital has been a controversial subject. This golden decade for the Chinese bourgeoisie represented in Communist historiography a collusion between Nanjing and the capitalists, especially those from Jiangsu and Zhejiang. Scholars, on the other hand, identified oppression by a ruthless regime serving interests of its own.[1] Recent studies, in contrast, emphasized the ad hoc nature and the institutional basis of that relationship.[2] For the salt industry, historians have focused on the successful efforts from a bureaucratic and institutional perspective of rationalization, professionalization, technocratic orientation, and buffering of the Salt Administration against external politicization under the new government, finding no contradiction between the laudable goals of modernization and revenue enhancement.[3] On the other hand, while contemporary critics continued to blame revenue farmers as appendages to a corrupt and bloated bureaucracy, Nanjing confronted the same problems as previous regimes had: finding a balance between revenue stability, industrial development, and people's livelihood.[4] For the refined salt industry, its relationship with the Nationalist government thus defies characterizations of collusion or collision; it was both.

This chapter analyses the vexing relationship between the salt industry and the new regime by tracing various attempts at reform: the National Fiscal Conference and the Salt Conference (1928), General Regulations on Refined Salt (1930), the New Salt Law (1931–1935), Re-registration (1933–34), revision of the General Regulations on Refined Salt (1934), and the Provisional Regulations on the Marketing of Refined Salt (1934–35). Contrary to some assertions, marketing of refined salt did not become free after 1932, nor were revenue farmers immune from rent-seeking officials.⁵ Jiuda and the refined salt industry drew on the resources of their network portfolio to navigate the contradictions between the new government's goals of promoting people's livelihood, industrial development, meeting its revenue needs, and the muddle of salt institutions, lobbying and cultivating relationships with key officials as well as legislators while orchestrating a campaign of public opinion. Development of the industry thus coexisted with bribery and corruption, by definition illegal and costly.⁶

The Revenue Imperative and the National Salt Conference

With the success of the Northern Expedition, Canton and Nanjing promised major reforms: Gu Yingfen pushed for the dissolution of the Revenue Inspectorate, while Song Ziwen declared revenue farming an anachronism on March 23, 1928. Of the dozens of proposals discussed during the National Economic Conference in June 1928, five dealt with salt gabelle and its reform, including abolition of revenue farming and "tax at depot and free trade thereafter" as guiding principles.⁷

Nevertheless, Fan Xudong and Jing Benbai agonized over the promised institutional change, drawing on their network for information and how to approach the new regime. The appointment of Xu Jingren (1872–1948), a leading Huainan crude salt producer, as chief commissioner of the Salt Administration confirmed their worst fears. A well-connected Zhou Zuomin reported that Xu and Qian Junda (a Harvard graduate) had set the new government's policy: To ensure fiscal stability, revenue farming and the divisional system must be preserved. Indeed, Nationalist leaders kept a close watch on salt and the gabelle, the bulk of which were paid by the revenue farmers, as a source for one-quarter to one-third of its total revenue.⁸ Insiders warned Jing and Fan that the new government was reluctant to allow any expansion of the refined salt industry in the provinces.⁹

This reluctance, however, should not be interpreted as collusion between the government and the revenue farmers. In practice, Nanjing also continued another feature of the salt trade: pay to play. Much has been made of the collusion between Shanghai's bankers with the young regime, obscuring the role of revenue farmers who secured the bank loans with their gabelle advances. To renew their legitimacy, Lianghuai revenue farmers bargained with the new government and raised, directly or indirectly, over 60 million yuan for Nanjing in advances on the gabelle, price increases, and bond purchases, including 3 million of the 30 million yuan treasury notes issued in May 1927, 60 million yuan in treasury notes secured by Lianghuai and Liangzhe gabelle surplus two months later, and another 3 million yuan loan secured on a price increase of one yuan per *dan* for military expenses in January 1928.[10] Where the merchants did not oblige quickly enough, as in the case of the five head merchants of Changlu, Jiang Jieshi ordered their arrest, brought them to Nanjing, and held them for over two years without trial until a fine of 500,000 yuan was paid.[11]

The relationship between Jiuda and the new regime began just as ominously. The company's location in Tanggu added to the uncertainties since a portion of its tax payments had been subsidizing, with the approval of the Revenue Inspectorate, the northern warlords. To address the inequity, the Nationalists imposed a "Northern expedition levy" and a special levy on refined salt (*jingyan teshui*) in transit through Shanghai. The once powerful Revenue Inspectorate, its very existence threatened by the Nationalists, "almost completely disappeared" in this "dark period."[12] With no recourse, the company again suspended operation until July 1928.

Drawing on structural holes in their network, Fan Xudong and Jing Benbai searched hard for approaches to the new regime. Zhou Zuomin was asked to discuss the matter of the special levy with Song Ziwen.[13] Through the American university old boy network, Cornell graduates Ren Shuyong (1886–1961), Yang Quan (1893–1933, style: Xingfo), and Hu Shi (1891–1962) brought the refined salt industry's predicament to the attention of Song and Zou Lin (1888–1984, Qian Junda's successor).[14] A mutual friend, Zhou Junyan, accompanied Jing to an hour-long meeting with Zou, but the Northern Expedition levy remained nonnegotiable.[15] Instead, Zou called a conference to discuss problems confronting the industry.

In the National Salt Conference from December 12 to 22, 1928, Jing found himself the lone representative of the refined salt industry among officials and revenue farmers. Of the 113 motions discussed, none dealt with the conflict between crude and refined salt, leading Chen Changheng (1888–1987), a Harvard-trained economist then serving as chief ministerial auditor and sympathetic to gabelle reform, to declare his time wasted.[16] The reason is clear: while the conference was in session, the ministry contracted a 5 million yuan advance from the Lianghuai and Liangzhe revenue farmers to help meet pressing end-of-year expenses.[17] Two months later, Song Ziwen, too, had a change of heart and began reregistering the revenue farmers of Jiangsu, Zhejiang, and the four Yangzi provinces. They were to present licenses issued under the Qing for verification with an examination fee (*yanzhengfei*) of one yuan per *dan* in exchange for new ones affirming their hereditary privilege in perpetuity. The merchants pleaded hardship and bargained, and the government settled with a 20 percent discount.[18] The new regime's pressing revenue needs and obliging merchants, not to mention allegations of bribery, thus complicated the prospect of reform and the future of the refined salt industry.[19] No mere appendages or victims of the Nationalist state, both the revenue farmers and the salt refineries lobbied and bargained hard over the proposed regulations and the laws governing the salt trade.

General Regulations on Refined Salt, 1930

With their legitimacy renewed, the revenue farmers attacked (see Appendix 9 for the market share of refined salt in the four Yangzi provinces, 1917–1934). Presenting themselves as staunch supporters of the new government, they became the leading advocates of state interests.[20] Accusations, many false, flew in their petitions: The refineries did not pay any reregistration fees (as if officials needed a reminder to rectify this oversight) or additional local levies (not true), nor were they subject to stringent and costly transportation regulations (ignoring, for instance, the gunny bag saga), or required to maintain a supply on hand.[21] Enjoying these unfair advantages, the refineries encroached on the revenue farmers' market, directly jeopardizing the state's revenue stability and the livelihood of salt producers.

Taking the accusations with a grain of salt, officials did little. Instead, they focused on raising more revenue. The Salt Administration queried the Lianghuai Inspectorate in 1930 about whether the revenue farmers and the refined salt industry might cooperate if the refineries were granted access to the division with a levy of one yuan per *dan*. The answer was noncommittal. Having just paid millions in reregistration fees, the revenue farmers could hardly be expected to cooperate. With the existing market for refined salt limited to Nanjing and Yangzhou, sales could not be expected to increase quickly in the short term. In addition, the extra levy would make refined salt even more expensive relative to crude salt, but the decision was Nanjing's alone.[22]

With no easy solution in reach, the Nationalist government chose to protect its revenue by promulgating laws on the inspection of refined salt (December 10, 1930) and the General Regulations on Refined

Salt (*Jingyan tongze*, December 31, 1930) that improved upon the ad hoc production and marketing rules accumulated since 1914.[23] For the first time, it was stipulated that refined salt must contain at least 95 percent of sodium chloride. Clause 2 of the General Regulations required a minimum registered capital of 100,000 yuan for each refinery, and official production capacity determined by how much capital a company possessed. Once approved, the registration permit would be valid for ten years, renewable six months before expiration (clause 11, with no mention of a registration or a renewal fee). Clause 13 required the refineries to report both their capital and planned production at the beginning of each year. An annual production of at least 30,000 *dan* was required, and the requisite gabelle guaranteed by a third party with liability for any shortfall (clause 14). Above all, clause 15 authorized the Finance Ministry to declare a condition of excess production capacity and restrict production accordingly. Supplemented by clause 35, which empowered the ministry to limit the quantity to be sold in any market, the new regulations imposed potential strangleholds on the refined salt industry.

The NSRB and the refineries fought back demanding fair and equal treatment. Crude salt did not have a minimum on tax paid or sodium chloride content. For the refineries to anticipate and file their production plans at the beginning of each year was a futile exercise of paper work as their production capacity was already on file with the ministry at the time of company registration anyway. There should also be a distinction made between registered capital, paid-up capital, and available capital (including bank loans). Setting production limits according to registered capital was thus a flawed method.[24] As a modern government promoting national health and industrial development, its arbitrary and unilateral power to limit production and marketing was unreasonable, if not self-contradictory. After all, the state did approve the registrations, collected the requisite fees, held the refineries accountable for the gabelle calculated on the basis of registered production capacity, and yet deny the refineries the opportunity to produce and market. Their pleas, however, fell on deaf ears.[25] Henceforth, production of refined salt, calculated at the rate of two yuan in capital for one *dan,* was capped at 2,490,000 *dan* for 1931, 2,415,000 for 1932, 2,245,000 *dan* for 1933, or approximately 6 percent of the country's annual consumption, while the Salt Administration's revenue guards and antismuggling patrols employed by revenue farmers were encouraged to stop leakages of refined salt from the commercial ports into the interior.

New Salt Law, 1931

While petitioning and forging ties to officials, Jing and Fan fought the new regulations with allies in the Legislative Yuan. Working with eleven sympathetic legislators, including leading fiscal experts such as Ma Yinchu (1882–1982), Chen Changheng, Wei Tingsheng (b. 1890, another Harvard-trained economist), Zhuang Songfu (1860–1940), and Liu Furou (1876–1953), Jing assisted in the drafting of New Salt Law with thirty-nine articles.[26] Invoking Sun Yat-sen's principle of people's livelihood, the proposed legislation promised to lower both price and tax while improving the quality of salt. Article 1 declared free trade of salt and abolition of all transportation and marketing monopolies. To qualify as alimental salt required at least 85 percent of sodium chloride by weight, with moisture limited to no more than 8 percent. Without identifying crude salt, Article 4 proclaimed that "salt consisting of less than 85% of sodium chloride shall not be used for alimentary purposes." Before delivery to consumers, all salt, whether crude or refined, must be stored in state depots after testing and verification. Producers would be required to reprocess any substandard salt. After paying a uniform gabelle of five yuan per *dan*, no local surtax or levy was permitted (Article 24). Article 38 spelled the doom of revenue farming. Introduction of the new law would nullify "all laws and orders basing upon the *yin* (i.e. licensed, mostly hereditary and some pro term scattered among the divisions) merchant, farming merchant, government transportation and sale and other similar arrangements." The proposed law emerged from committee on March 21, 1931, intact, surviving a last minute attempt by Song Ziwen to revise the draft.[27]

Indeed, Song and his subordinates at the ministry, including the foreign staff at the restructured Salt Administration, faulted the proposed legislation for a variety of reasons. Even Chen Changheng conceded that implementing the law required construction of state depots at great expense while planned production and stringent quality control necessitated closure of inefficient facilities with significant impact on merchants and thousands of salt producers. Ideally, free trade rendered the expensive antismuggling forces obsolete and reduced government expense, but there was no guarantee, given the long history of salt smuggling, that expensive relief and welfare programs for unemployed salt producers would work. Even more troubling to Song was that the uniform five yuan gabelle might not cover the necessary construction projects and programs for displaced salt producers with enough left to help meet Jiang Jieshi's military expenses.[28]

F. A. Cleveland, the first American to serve as associate chief inspector of the reorganized Salt Administration, similarly raised grave concerns in his analysis. Bringing under control "all the areas where salt may be produced and all the works" with a new system to organize and finance the production, handling, and distribution of salt on a free trade basis, including the most remote corners of the country were laudable goals. But he also doubted that the revenue farming system—"a highly complex but well organized and highly capitalized system"—could be abolished by legislative fiat without any compensation to those who had paid the "reregistration" fees.[29] Even more pessimistic, Wolcott saw no hope for an inspectorate reduced to a purely advisory role. He had only contempt for a "...brainy, but as crooked as a hunch-backed snake" Song Ziwen and other Nationalist "money and fame-grabbers."[30]

Complicating the debate even more, a leaked draft of the new law sparked an intense lobbying campaign in a battle of cables, petitions, and editorials, paid or unpaid.[31] Before the vote, a mercurial Zhang Hu reportedly arrived in Nanjing to work his magic for the revenue farmers. On behalf of Liangzhe revenue farmers, Zhou Qingyun (a major shareholder of Wuhe as well) offered Song Ziwen emendations to the draft, including a gabelle rate increase to eight yuan per *dan*, which the legislators rejected.[32] Over dinner with the drafting committee, Jiang Jieshi failed to break the deadlock between the legislative and executive branches. On May 12, 1931, the People's Political Council passed the New Salt Law and ordered the Finance Ministry to form a commission within three months to design an implementation plan. There was, however, a loophole: Article 39 of the New Salt Law specified that "the date of introduction of this law shall be decided by separate orders."

Those orders never came. One explanation for the stalled effort focused on the bargaining between the revenue farmers and Nanjing government. In exchange for underwriting another 80 million yuan in short-term treasury bonds secured on the gabelle, the Nationalists allegedly assured the revenue farmers that the new law would not be introduced until the bonds were repaid in full.[33] Ma Yinchu attributed the cause to obstruction by revenue farmers.[34] Work on the new law, however, did continue with the Organic Law of the Commission for Salt Administration Reform promulgated on May 2, 1932, followed by an intense lobbying campaign by both the revenue farmers and the refined salt industry to pack the commission.[35]

Implementation of the New Salt Law also lagged because of the dilemmas confronting Nanjing. Critics of the law did raise significant concerns. Licenses inherited from the Qing dynasty might be of dubious legality, and the revenue farmers should have paid even more gabelle because of their cheating, wastage allowances, and a revenue quota set centuries ago. However, to pay for the reregistration fees and operations, revenue farmers did amass huge bank loans that must be repaid, and their failure to repay would compromise the government's credibility.[36] With free trade also came uncertainty about responsibility for the gabelle, or supply of salt to remote corners of the country. The ideal of a uniform gabelle rate, improved quality of salt for the masses, tax and price rollbacks (if any) through free trade, and increased efficiency must be weighed against the realities of revenue shortfall and widely varying production costs. Free trade could drive saltyards in the interior with high cost out of business, leaving only Changlu, Shandong, and others on the coast.[37] Indeed, when word spread that Sichuan was to experiment with the first stage of free trade, that is, saltyards would not enjoy the privilege of being assigned specific and exclusive markets, producers in northern Sichuan mounted a fierce protest campaign. With their costs reportedly twice as much as those in southern Sichuan, they would be doomed. Charged with implementing the reform, even Miao Qiujie had second thoughts. With Japanese invasion threatening the coastal salt fields, strategic concerns, too, must be considered.[38] Modern China's salt morass defied a simple solution.

Reregistering the Salt Refineries

Under siege by the revenue farmers and weighed down by stringent new regulations, the salt refineries survived through a costly reregistration process, turning the crisis into an opportunity.

The campaign began for Jiuda on January 10, 1933. As the Japanese invaded and threatened Northeastern Hebei, Yu Xuezhong, then governor of Hebei, needed to find ways to meet pressing military expenses. Working through Changlu commissioner Jing Youyan, another member of Zhang Xueliang's inner circle, Changlu revenue farmers were ordered to present their licenses from the Qing dynasty for verification. Citing the Lianghuai and Liangzhe precedent of 1929, they, too, must pay a reregistration fee of one yuan per *dan* on their quota within a month. On January 22, 1933, the commissioner belatedly amended his order to include refineries under his jurisdiction, that is, Jiuda and Tongda. The companies faced a bill calculated on the basis of 0.5 yuan per *dan* of their averaged sales in 1931 and 1932.[39]

An old hand at negotiating with officials, Jing stalled by raising issues that even seasoned secretaries at the Changlu commissioner's office found difficult to dismiss.[40] Enjoying neither hereditary privilege nor an assigned exclusive market by the authority of antique Qing licenses, salt refineries were governed by a separate and unequal set of regulations with no such licenses to examine. Another issue concerned the consistent application of the law: if Changlu merchants somewhat belatedly must pay because of precedents set elsewhere, should refineries in other divisions be reregistered and pay the fees as well? If not, how could Jiuda continue in business under such a handicap?

While the war of words raged in Tianjin, Zhong Lijian, NSRB's lobbyist in Nanjing, reported what he learned from Ma Taijun (b. 1879, a native of Tianjin, then chief of the Salt Administration's marketing bureau and a classmate of Zhu Tingqi at Peiyang University and Harvard) and Zeng Yangfeng (b. 1886, a graduate of the University of Illinois and, as head of the production bureau, one of the four "guardian gods" of the Salt Administration). Yu Xuezhong had engineered the campaign with Jing Youyan, hoping to raise at least 2 million yuan to meet Zhang Xueliang's pressing needs. Ma and Zeng both doubted the validity of the orders, but there was little that Nanjing could do. As a matter of strategy, Zeng candidly advised Jiuda to pay for the 8,000 *dan* of shipments within Changlu but stall on the rest.[41]

Unaware of Ma and Zeng's advice, Jing Youyan pressed his demands. He did not dwell on the issues raised in Jiuda's petitions, claiming instead that at a time of national crisis, the company should put the country ahead of its word-mincing. Having profited for years from the state's generosity, it should pay forthwith or risk suspension. Despite the threat, the company remained defiant. The commissioner might disregard the law in such difficult times, and Jiuda would have complied except it did not understand what it was paying for: reregistrating its license as a revenue farmer with monopolistic privilege; or access to a share of the Changlu market with a fee which was not assessed on refineries in other divisions; or a donation to the state.[42]

While Jing Benbai negotiated with Jing Youyan through Ye Pinghou (an official dispatched from Nanjing to facilitate a solution) and Cai Guoqi, Fan Xudong drew on his expanding network within the Nationalist government. Since his futile negotiations in 1929 with Nanjing over a state capital infusion into Yongli and his appointment in 1931 as a member of the National Fiscal Committee (*Caizheng weiyuanhui*) and the Ministry of Industries' Nitrogen Fertilizer

Committee, he had worked with senior officials in Nanjing. Aware that Song Ziwen paid scant attention to telegrams in Chinese, Fan cabled Song in English on January 28, 1933: "Even for the purpose of raising money the amount so indicated will be very small indeed. It seems it will not pay the Central government so to break the usual laws if the amount involved is only so insignificant." He suggested convening a national conference of salt refineries to raise a larger sum in exchange for the "removal of their unnecessary bond such as the limitation of consumption territory etc...(to) avail the people of an opportunity to select and purchase salt of better quality..."[43]

Whether it was due to his public-mindedness or a strategy of paying to play, or both, Fan's suggestion also killed several birds with one stone. All the bickering and games among the refineries over market shares could ease with an enlarged market.[44] At the very least, the move diverted some of the pressure and attention away from Jiuda. Intrigued by the suggestion, Song Ziwen instructed his Harvard schoolmate Zhu Tingqi (b. 1888, successor to Zou Lin in 1932) to pursue the possibility.[45] Three days later, Fan arrived in Nanjing.

Untangling the mess took several months of tortuous negotiations and lobbying in Nanjing, Beijing, and Tianjin. Venting his frustration with the company, Jing Youyan ordered the suspension of Jiuda's shipping permits on February 11, 1933. Fan countered the next day by offering Zhu Tingqi a gabelle advance of 100,000 yuan in cash, an offer which secured a Salt Administration cable staying Jing Youyan's order until a plan to reregister all the refineries could be formulated.[46] After meeting with Zhou Zuomin, Song Ziwen summoned Jing to Beijing to work with Cai Guoqi to break the impasse. However, Jing continued his threats. After another round of negotiations pitting bureaucrats against each other, a cash payment of 3,000 yuan and another 107,840 yuan of advance gabelle in promissory notes settled Jiuda's troubles with the Changlu commissioner.[47]

Reregistration of the refineries, however, took much longer. Fan Xudong and Zhong Lijian persuaded Zeng Yangfeng, Ma Taijun, and Zhu Tingqi that, for the purpose of raising revenue quickly for the central government, the refineries might be able to afford a fee of 0.5 yuan per *dan* assessed on 1.3 million *dan* (average sales of refined salt 1930–1932: 1,198,751 *dan*), or 650,000 yuan. In exchange, the Salt Administration promised that only refineries that had paid the fee could participate in the trade and a moratorium on new refineries for a period of five years. Although costly, Fan and his colleagues calculated that paying this fee might secure the refineries a state-sanctioned share of the salt market with access to the interior, if not equal treatment to privileged revenue farmers.[48] Remaining major issues concerned the Yingkou refineries, which had fallen under Japanese control, and what to do with their share of the market. Zeng agreed to ban further shipments from Yingkou. Fan projected Jiuda's sales to increase by one-third to half a million *dan*, a prize worth paying for, either as part of the resistance against Japanese invasion or the company's business.

Negotiations with the Nationalist bureaucratic labyrinth, however, could be daunting. Issued while Zeng Yangfeng was on vacation, the Finance Ministry's order 6252 to NSRB on March 31, 1933, contained surprises. The carefully crafted order specified posting Nationalist officials at all the refineries to supervise production and shipment, a requirement that the Japanese Kwantung army rejected, resulting in a de facto embargo against Yingkou shipments on April 5, 1933. On the other hand, Song demanded a registration fee of one yuan per *dan* and reduced the moratorium on new refineries from five years to three. The order also specified that the 1.3 million *dan* should be divided among the eligible refineries on the basis of their average sales between 1930 and 1932.

Extant archival sources are silent on why Song overruled his subordinates. Perhaps he wanted to maximize Nanjing's revenue at a time when its monthly deficit exceeded 10 million yuan, or other lobbyists had his ear, or both. On behalf of Jiuda, Zhou Zuomin again cabled, but found Song preoccupied with preparations for his world tour.[49] Whatever the reason, the order touched off another round of negotiations and lobbying. By the ministry's calculation, Jiuda would only be alloted 38.65 percent of the market on sales of 288,780 *dan* (see Table 6.1). Working his magic again, Zhong Lijian reported to Fan that a sympathetic Zeng Yangfeng promised to rectify the problem. Jiuda should petition the government that its business had been hurt by the pooled sales agreements and political chaos. With the largest capital and state-approved production capacity in the industry by far, combining sales and capacity was a more equitable formula. Jiuda would thus command 44.75 percent of the market as opposed to Tongyi's 25.1 percent.[50] Zhong and Fan also prevailed upon Zhu and Ma to approve a 20 percent discount on the registration fee, in line with what most Lianghuai revenue farmers paid. Satisfied with those terms, Fan pleaded with Jing Benbai: the industry should not stall any more in this time of national crisis.[51]

Persuading the other refineries to follow Fan's lead proved difficult, however. With the Yingkou refineries absent, NSRB met on

Table 6.1 NSRB calculations of registration allotment (unit = *sima dan*)

Company	Production Cap	3-year Average Sales	% of Sales	Average Sales Only	Average of Production/Sales	% of Total
Jiuda	1,050,000	288,780	38.65	502,450	581,750	44.75
Tongyi	500,000	251,100	33.6	436,930	326,300	25.1
Yongyu	400,000	96,880	12.96	168,610	215,930	16.61
Wuhe	120,000	65,004	8.7	113,100	80,340	6.18
Tongda	50,000	24,304	3.25	42,380	32,240	2.48
Minsheng	50,000	18,964	2.54	32,890	29,900	2.30
Dinghe	75,000	2,100	0.28	3,640	32,540	2.58
Total	2,245,000	747,132	100	1,300,000	1,300,000	100

May 18, 1933, in Shanghai and found much disagreement. Tongyi supported Song's formula, which gave it 33.6 percent of the market. The smaller refineries, led by Dinghe, which began production only in 1932, argued that the government's order was unfair and demanded a larger allotment.[52] With no concensus, the meeting adjourned with an agreement to proceed as each saw it fit.[53]

Jiuda's company records offer only a glimpse of the lobbying battle. One day before the bureau chiefs' recommendation to Zhu Tingqi was due, a surprised Zhong Lijian uncovered a scheme awarding a 20 percent bonus to refineries consistently attaining at least half of their combined sales and production capacity average. Reportedly inserted by Zeng Yangfeng at the instigation of Lin Ziyou, Tongyi's share of the market would increase by 5 percent at the expense of Jiuda and Yongyu (see Table 6.2). Separately, Dinghe also found an advocate in Zou Lin (vice-minister of finance). To protect smaller refineries from "predatory" big ones, any refinery with a production capacity under 100,000 *dan* would be entitled to sell as much as it could up to its registered limit. In a final push, Zhong tried unsuccessfully to exploit the disagreements between the production and marketing bureau to have Zeng's bonus scheme suppressed, although Ma Taijun promised a personal letter to Zhu Tingqi in support of Jiuda's case.[54]

The reregistration order issued by Zou Lin in Song Ziwen's name on June 5, 1933, was a product of the lobbying efforts. As shown in Table 6.2, the four smaller refineries were awarded a total of 295,000 *dan*. Wuhe, a company that had been dormant since 1932, saw its share increase by 3 percent to 9.2 percent, or 120,000 *dan*.[55] Jiuda, Tongyi, and Yongyu were to apportion the remaining 1,005,000 *dan* according to their averaged three-year sales record and production

Table 6.2 State-sanctioned registration/market share, June 5, 1933

Company	Average of Production Cap 3-year Average Sales)	% of Total (with 20% bonus)	State-approved Share	% of Total
Jiuda	534,300	41.4	520,590	40.00
Tongyi	391,300	30.1	291,450	22.4
Yongyu	179,600	15.2	192,960	14.8
Wuhe	80,600	6.2	120,000	9.2
Tongda	32,500	2.5	50,000	3.9
Minsheng	29,900	2.3	50,000	3.9
Dinghe	33,800	2.6	75,000	5.8
Total			1,300,000	100

capacity at a ratio of 51.8: 29: 19.2, respectively. As a result, Tongyi's market share shrunk by almost 8 percent. Lin Ziyou conceded defeat in an angry note: what concessions that Fan could not obtain in the NSRB negotiations he won through officials. A bank syndicate, led by Jincheng, underwrote Jiuda and Yongyu's reregistration with a 500,000 yuan loan, repayable in full in two years at a monthly interest rate of 10 percent.[56]

Nanjing thus emerged as the biggest winner in this episode. By September, Jing Youyan was replaced, despite the windfall he generated for the central government, by Wang Zhanghu (b. 1878) as head of the reorganized Changlu division. Instead of 145,933 yuan as originally demanded by Jing (on an average of 291,867 *dan* shipped outside of Changlu between 1931 and 1932), Jiuda paid over twice as much in reregistration fee to Nanjing. Collectively, the industry paid a reregistration fee of 1,040,000 yuan at 0.8 yuan per *dan*, a 20 percent discount on Song Ziwen's original demand but in cash while revenue farmers paid by installments.[57] Preoccupied with capturing their registration allotments, the refineries also overlooked the fact that their attempt to secure equal treatment with the revenue farmers remained unfulfilled. Even worse, sale of refined salt beyond the "commercial ports" was still illegal.

The lobbying battle thus continued. On July 16, 1933, Zhong reported another reversal: On the pretext that Tongda, Dinghe and Minsheng had yet to pay their reregistration fees in full, the Finance Ministry reneged and decided to lift the embargo against the Yingkou refineries. The Northeastern provinces might have fallen, but the companies were Chinese with 400,000 *dan* of inventory that must not fall into enemy hands. An outraged Lin Ziyou immediately demanded a refund of the reregistration fees he paid, while Fan Xudong cabled Zhu

Tingqi reminding him of his promises. Subsequently, Jiuda's sources uncovered the who, what, and how much: the Yingkou refineries as well as their creditors, including the Bank of China, allegedly secured the reversal through Zhang Shouyong (1876–1945), until recently vice minister at the ministry. In addition to the regular gabelle, they offered a "patriotic duty" (*aiguo juan*) of 120,000 yuan, as well as 80,000 yuan for Zhang and Zou Lin's services.[58]

Candid conversations between Zhong Lijian and Ma Taijun revealed disagreements among the officials. Some opposed any deal with the Yingkou refineries, but the ministry needed a minimum of 5 million yuan each month to keep the government afloat.[59] As a compromise, Bei Gongyan (b. 1887), a Brown-trained section chief proposed charging the Yingkou refineries a levy of 1.6 yuan, half of which would compensate the refineries that had already paid the reregistration fee, and half to the state. The three-year moratorium on new refineries would be restored to five years. On the other hand, the Yingkou refineries insisted that as a one-time arrangement, they should not be assessed any reregistration fee, or at least not until their relocation to Huaibei.[60] Overruling Zhu Tingqi, the ministry lifted the embargo but with strings attached. They must, in addition to the 1.2 million yuan of gabelle, pay a "patriotic duty" of 200,000 yuan. Compensating the other refineries for the reregistration fee was a matter for the merchants to settle among themselves.[61] Unable to meet those demands, the Yingkou refineries declared bankruptcy, their inventory purchased by the puppet regime, and salt pans absorbed by the Manshū engyō kabushiki kaisha [Manchurian salt industries, Ltd.][62]

However, even the removal of the Yingkou refineries did not improve the market for refined salt. In Hankou, the largest market, sales actually declined because of flooding and competition from crude salt from Huaibei and gypsum salt from Yingcheng, as shown in Table 6.3.[63]

Table 6.3 Sales of refined salt in Hankou, 1932–1936 (unit=*dan*)

Company/Year	1932	1933	1934	1935	1936
Jiuda	107,250	114,240	109,200	119,254	115,182
Yongyu	45,360	70,560	57,178	52,704	N.A.
Tongyi	N.A.	90,000	110,490	98,942	N.A.
Wuhe	N.A.	10,080	29,870	27,674	N.A.
Minsheng	N.A.	8,736	9,601	20,396	N.A.
Tongda	N.A.	14,784	27,505	25,884	N.A.
Total	434,046	366,023	373,328	342,714	346,000

Nationally, sales of refined salt in 1933 (1,204,989 *dan*) and 1934 (1,152,717 *dan*) fell short of the 1.3 million *dan* limit. To reverse a shrinking market, the refineries mounted a sophisticated campaign to break out of the "commercial ports."

Revising the Regulations, 1934–5

The campaign began with a cable from Zhong Lijian to Fan Xudong on December 24, 1933, reporting a conversation he had that day with Zhu Tingqi. The chief commissioner reaffirmed his personal support of the refined salt industry with a proposal to mandate conversion of Lianghuai crude salt production to refined salt. Zhong reminded the chief commissioner of one precondition: refined salt must be permitted access to the interior as well. Zhu reiterated that he expected strong opposition, his rivals at the ministry already accusing him of being partial to the refined salt industry, and he needed support. Perhaps, Zhu suggested, Fan Xudong might help convince Kong Xiangxi (1880–1967), the minister.[64]

That Zhu would solicit a private citizen's aid in a fight among bureaucrats reflected Fan Xudong's rising stature in the national economy and his expanding network. Taking as his duty to build the country's first chemical fertilizer plant in Nanjing without any direct state investment, he served as a charter member of the National Defense Planning Committee (Guofang sheji weiyuanhui, precursor to the National Resources Commission) with direct access to the highest echelons of power in the government, including Jiang Jieshi.[65] Encouraged by Zhu Tingqi's overture, Fan Xudong and his colleagues mounted a campaign mobilizing the Legislative Yuan, contacts within the Nationalist Party and government, and a media blitz, culminating in a petition drive to revise the General Regulations on Refined Salt.

The campaign opened with the NSRB and various chambers of commerce petitioning the Legislative Yuan and the Political Council legislative committee. Working closely with Chen Changheng, chair of the economic subcommittee, and Shi Shangkuan (1898–1970), chair of the review subcommittee, Zhong drafted briefings to help them build a case for refined salt.[66] The objectives were simple: revision of the regulations to remove as much as possible restrictions on the industry, especially Articles 33 and 37, which confined consumption and marketing of refined salt to the "commercial ports." After days of deliberations and debate, the committees voted to allow not merely access to treaty ports and ports declared open for trade, but also any area designated for free trade of salt. Even more importantly, consumers were allowed to purchase and transport refined salt beyond the ports.[67] Approved by the full committee on June 16, 1934, the revised regulations were forwarded to the Political Council for review. However, amid a constitutional convention and hundreds of other pending legislation, the revised General Regulations on Refined Salt languished in council beyond the reach of even the legislators or the lobbyists.

The refineries could not afford to wait, however. Squeezed by the vigilant revenue guards of the Salt Administration, revenue farmers' private antismuggling patrols, and a poor market, sales continued to decline.[68] With half the year gone, shipment had reached only a quarter of the registered limit and the banks grew increasingly anxious about repayment of their loans.

Again, pressing need for revenue created the opportunity that the industry exploited. He Jian, governor of Hunan, had demanded an advance from salt merchants to finance another extermination campaign against the Communists. When the revenue farmers balked this time, the refineries risked 800,000 yuan in return for a provincial order permitting consumers beyond the commercial ports to purchase and transport up to eight bags of refined salt, citing the precedent of Hubei.[69] An offended Kong Xiangxi protested against this violation of the law and the central government's authority to set gabelle policy, but neither the Salt Administration nor the Finance Ministry had the cash to pay He, or the power to overrule Jiang Jieshi.[70]

Capitalizing on this fait accompli and his discussion of the industry's plight with a seemingly sympathetic Kong Xiangxi at the National Defense Planning Committee meeting of August 1934 in Lushan, an optimistic Fan and his colleagues redoubled their lobbying effort. The industry had paid its gabelle and reregistration fees as demanded, yet was denied equal treatment and protection under the existing regulations. Instead, harassment continued. Humbly, they begged the government for relief.[71] Behind the scenes, Zhong Lijian coordinated the drive with growing sophistication. Taking advantage of the convocation of the Nationalist Party's fifth central executive committee meeting (December 10–14, 1934), he coordinated an elaborate campaign. Cables from chambers of commerce and various local chapters of the NSRB poured into Nanjing where their delegates debated officials of the Salt Administration and the ministry, supplemented by radio broadcasts, press conferences, and press releases.[72]

The well-timed campaign, reinforced by Fan Xudong's cables to Zhu Tingqi imploring him to keep his promises, prompted action.[73]

Zhu summoned Zeng Yangfang and Ma Taijun to Shanghai to discuss the matter, but found their proposals wanting. The bureau chiefs recommended that no more than ten *liang* (16 *liang* = 1 *jin*) of refined salt per person be carried beyond the "commercial ports." Zhu had in mind fifty *jin* per person per instance. Late that evening, Zhong Lijian received a call summoning him to another meeting the next day.[74]

Catching the overnight train, Zhong arrived at the Salt Administration's Shanghai Office to find a new negotiator. On temporary leave to attend his mother's burial, Zhu Tingqi had assigned Chen Tianji (style: Gongliang, b. 1894; a graduate of Lehigh University who became chief of the Revenue Guards Department on August 29, 1934; and a Yongli shareholder with fifteen shares) to handle the negotiations in the hope that Chen, fresh to the economics and politics of the Salt Administration and the ministry, might be able to craft a solution.[75]

Both natives of Zhejiang (Zhong from Yuhang, and Chen from neighboring Haiyan), the two collaborated well. Chen accepted Zhong's briefing papers and agreed to meet with a delegation of NSRB local chapters. Together, they crafted a mutually acceptable proposal and presented it to Zhu on November 25, 1934.[76] Refined salt, with no gabelle-free wastage (*luhao*) and weight allowance (*jinyu*), generated more gabelle revenue by weight than crude salt for the state. Offering consumers a cheap and hygienic product satisfied the people's freedom of choice. As a preemptive concession, the refineries offered to limit sales permanently to the registered production capacity in exchange for immediate access to the interior beyond the commercial ports. Merely affirming a fait accompli, the revenue farmers' opposition was rendered moot and the lawsuits against revenue guards and antismuggling patrols cleared. Two days later, the trio met to discuss two further options: open access to all divisions with a fee of 0.5 yuan per *dan*, or lifting all marketing restrictions beyond the commercial ports with sales limited to an average of the past three years. Zhong convinced Zhu that the refined salt industry, on the brink of bankruptcy with a slim profit margin, could not afford the fee.[77] Finally, Chen was directed to draft the Provisional Regulations on the Marketing of Refined Salt.

The resulting draft improved slightly on Zhong's proposal. Sales of refined salt beyond the "commercial ports" in the four Yangzi provinces and the five prefectures of Suzhou were limited to the average sales in the past three years up to the registered 1.3 million *dan*. For the time being, the remaining 945,000 *dan* of authorized production capacity could be marketed in other parts of the country designated for free trade. Seizing the momentum, Zhong persuaded Zhu to make the draft official immediately.[78] At Zhong's urging, Fan also wired the next day to reinforce Zhu's resolve:

> 447 Rue Lafayette Shanghai Personal stop understand refined salt companies are in very desperate condition owing to so far only one quarter of authorized quantity sold stop From standpoint of registered amount the quantity sold is only one half stop as year is soon to close banks are beginning to press hard for repayment stop need of immediate and fundamental relief most necessary stop glad to hear you are taking measures to that end hope you will act speedily.[79]

On behalf of Fan, Zhong promised that no effort would be spared to persuade Kong Xiangxi. Overcoming the opposition of Miao Qiujie and others, Zhu persevered with Chen Tianji's determined support.[80] The proposal was formally submitted to the ministry on December 7, 1934.

As an official document, the proposal was refreshingly candid.[81] Regulations restricting refined salt to the commercial ports had been a dead letter since the passage of exemptions in Hubei, Jiangxi, and most recently, Hunan. Instead of denying the reality while being swamped by lawsuits, the government proposed to grant the industry access to the interior within the registered 1.3 million *dan* limit, distributed among the various markets according to sales in the past three years. Unused limit would be transferrable from year to year. Revenue farmers should not object as this merely affirmed the status quo. The difference between the sales limit and production limit, or 945,000 *dan*, could be sold in other parts of the country declared free for salt trade, including Nanjing, a source of much trouble for the industry, the revenue farmers, and the revenue guards.[82]

Fan Xudong also kept his side of the bargain. At his request, Zou Bingwen (1893–1985; Cornell BSc., then assistant general manager of the Shanghai Commercial and Savings Bank) met with Kong on December 8, 1934, to plead the industry's case. When Kong complained that he had performed many favors for Jiuda and Yongli, Zou judiciously disclosed that Fan, as a token of his appreciation, had contributed 5,000 yuan to the Chinese Cultural Foundation (*Zhonghua wenhua jijinhui*) in Kong's name.[83] At Zhu Tingqi's suggestion, Zhong Lijian also approached Li Yuwan (style: Qingxuan, 1897–1978), Kong's influential personal secretary. The Provisional Regulations on the Marketing of Refined Salt became law on December 22, 1934.

The price of victory came steep. The petition and media drive cost the refineries at least 20,000 yuan (although revenue farmers charged that several hundred thousand yuan had been spent), ten yuan for each favorable report, and outspoken newspapers such as Gong Debei's *Jiuguo ribao* [*National Salvation Daily*] retained for 2,000 yuan.[84] In addition to the 5,000 yuan donation in Kong's name, the Chinese Cultural Foundation also received another 1,000 yuan from Jiuda as solicited by Chen Changheng. Against Fan Xudong's specific instructions not to get involved other than as a witness, Zhong Lijian was obliged to guarantee personally an 80,000 yuan loan from Jincheng Bank to the Dingji real estate cooperative [*Dingji fangchan hezuoshe*] run by Shi Shangkuan or risk the legislator's support. Cai Guoqi, Ye Pinghou, and Chen Tianji all received unspecified gifts as a gesture of goodwill after they discharged their official duties.[85]

The Nanjing decade was thus a golden period to the Nationalist government and officials, but less so for the refined salt industry or Jiuda. Although refined salt legally gained a toehold on the market beyond the commercial ports, sales never reached the 1.8 million *dan* record set in 1930 despite a sympathetic legislature, the principle of people's livelihood, and changes in regulations (see Appendix 8 for the state-set market share for the refineries in 1935). Reforming the gabelle system and improving the supply of alimental salt to the country proved equally frustrating. For all its efforts, the industry had had to settle for at most 2.245 million *dan* in sales (the limit of its production capacity) each year, or a little over 6 percent of the country's salt market.

An irrational, if not feudal, institution such as revenue farming was not the only barrier against the refined salt industry, however. Even a modern and professional bureaucracy such as the Revenue Inspectorate counseled that, to maintain revenue stability, free trade should wait. The revenue needs of the Nationalist government in a decade of state-building and continual crises made reform difficult. However regressive the gabelle, it remained a stable source of revenue collected through a well-capitalized farming system. Between 1932 and 1937, four rate adjustments created a complex tax table containing eighty grades ranging from 0.4 to 10.40 per *dan*. During the same period, the national average gabelle rate, not including various local and provincial levies, more than doubled from 2.83 to 5.98 yuan per *dan*.[86] Reform to promote gabelle uniformity by lowering and raising rates—and more areas saw increases (58) than rollbacks (25) —also meant that the proportion of gabelle revenue from low rate areas (below one yuan to three yuan per *dan*) declined from 35 percent in 1930 to

25 percent in 1937, while high rate areas (over six yuan per *dan*) grew from 19 percent to 43 percent during the same period.[87] Revenue thus almost doubled from 119,638,000 yuan to 217,705,000 yuan (1937) on essentially flat sales of 42,117,000 *dan* (1927) to 40,366,000 *dan* (1937).[88] The tax on salt accounted for between 20 and 30 percent of the Nationalist government's revenue during the period.

Then there were revenue farmers and license holders ever ready to oblige officials and the government in exchange for their renewed privileges. Reregistration fees, bond purchases, and promissory notes with payment guaranteed by issuing banks, not to mention the array of customary subsidies and gifts to local officials continued to work their magic during the Nanjing decade. Although the refineries, too, paid their dues, their pockets were not as deep.[89] As Jing Benbai succinctly put it, officials found it difficult to bite the hand that fed them.

Such was the nature of modern China's salt market and institutions that Fan Xudong and his colleagues found managing their portfolio of network resources a crucial part of their work. To survive, the refineries, individually and collectively through the NSRB, lobbied ministers, their secretaries, and bureau chiefs, exploited bureaucratic rivalries and divisions among the provincial and central governments, and manufactured public opinion. Costly as it might be, cultivating relationships with officials such as Kong Xiangxi, Zhu Tingqi, and Chen Tianqi finally secured for refined salt legal access to the interior.[90] Neither collusion nor collision described their complex relationships.

7
At War

On July 7, 1937, Fan Xudong left Tianjin for Nanjing on business. He never returned, for late that night the Japanese military launched its full-scale invasion of China. Writing to Hu Shi, then Chinese ambassador to the United States, Fan was defiant:

> China suffered greatly in the past year, yet our resistance remained resolute. Although I have lost over thirty million yuan and all that which took me two decades to build, I absolutely have no regrets. Indeed, I welcome this challenge, for only thus can China change...[1]

The country did change, but not entirely in the direction that Fan had hoped. Thousands of companies and industrial enterprises chose to stay in occupied territories or to seek refuge and profits in the foreign concessions, while many business elites in North China "actively collaborated with the Japanese from the beginning."[2] Jiuda-Yongli, the only major conglomerate from North China, evacuated with 600 other industrial enterprises from Lower Yangtze (354 with state financing) to the interior where they found conditions difficult and faced an increasingly nationalized economy.[3] Although Fan again drew on his network to serve the country's needs, officials attempted to annex the group during the war and moved to seize Jiuda as enemy property after the war, blurring further that porous line between collaboration and resistance.[4] Patriotism did not pay.

This chapter analyzes Fan's patriotic capitalism by tracing the relationship between Jiuda and the Nationalist state during the Second World War. After paying more than he could have had in reregistration fees and answering the government's call to advance over 10 million yuan in gabelle for the creation of a strategic reserve (ever-normal salt), he chose to join the war effort to increase salt supply in the interior. Continuing his commitment toward industrial social welfare, he relocated his crucial engineers, workers, managers, and their families as a unit to Sichuan.[5] Remaining true to his mission of national salvation through technology and industry, Fan shared his expertise and innovations at the expense of his profits. However, as he continued to promote free trade of salt and private enterprise in industrial development in the interior, his third path ran afoul of bureaucratic turf wars and the drive toward a state-controlled economy through nationalization.

Evacuation

As war closed in, Jiuda retained Otani Harusuke as an adviser on August 6, 1937 in the vain hope of buying some time.[6] The next day, Tanggu fell and Japanese soldiers surrounded the company, demanding its reorganization as a Sino-Japanese joint venture. After Jiang Jieshi declared his determination to resist at Lushan, Fan, who also attended the meeting, sought assistance from the government: he would rather mourn his loss than collaborate.[7] Persuaded by the vital role of Fan's group in national defense, Qian Changzhao secured Jiang's approval for a 3 million yuan subsidy to finance relocation of Fan's group to the interior and compensate for its loss, a most generous offer considering that over 100 Shanghai factories had had to share a 5 million yuan budget.[8] Fan instructed Li Zhuchen at Tanggu to dismantle critical machinery and evacuate some 300 key personnel, blueprints, and what equipment he could salvage to Sichuan. The Japanese military seized what was left.

By the time Fan and his colleagues made their way to Sichuan in February 1938, they found an old adversary, Miao Qiujie, in charge as the provincial Salt Commissioner. The technocrat had faithfully executed the state's policy of restricting Jiuda's expansion beyond the commercial ports to ensure the gabelle flow from Lianghuai revenue farmers. Transferred in 1936 to solve Sichuan's overproduction problem, Miao and his superiors reversed course as the salt-producing coastal provinces fell to the Japanese.[9] To stimulate production, the government provided loans and subsidies to Sichuan merchants to reopen dormant brine wells and saltworks to serve a vastly expanded market as the redrawn divisional boundaries gave them exclusive access to Sichuan, Hunan, Hubei, Henan, and Shaanxi.[10]

Unlike other evacuees, Fan and his colleagues thus found a hospitable host in Miao who personally conducted for them a tour of the various saltyards.[11] Drawing from these visits, Fan identified three problems in implementing the state's new objectives: high costs stemming from a limited brine supply, inefficient energy use for evaporating salt, and transportation difficulties. As part of his comprehensive plan to industrialize the southwest, he proposed the creation of a model salt factory utilizing the company's engineering expertise to improve the quality of salt, reduce cost by 25 percent, and increase production.[12] One week later, Fan cabled Miao that Kong Xiangxi had endorsed his proposal.[13] On April 4, 1938, Fan hand-delivered a detailed plan to Miao for a plant with a designed annual production capacity of 1 million *dan*, and a promise to share the company's technology with other salt producers.[14] The Salt Administration approved the plan quickly, directing Fan to take over a Zigong salt merchant-financed plant at Zhangjiaba that had been under construction since 1935.[15] Two days later, Fan's group finally received, after much delay and a concerted effort by Zhang Qun, Lu Zuofu, and Weng Wenhao, the first installment of 400,000 yuan of the 3 million promised subsidy.[16] Jiuda assumed control of the site on April 19, 1938, and construction began a week later with equipment evacuated from its Dapu plant.[17]

According to one longtime Jiuda employee, never before had the company's dealing with the government been so smooth.[18]

Troubles

This convergence of state and private interests brought on by the outbreak of war did not last, however. Sichuan salt producers and merchants opposed to free trade, the need to preserve the existing gabelle system (and thus revenue), brine and coal shortages, and bureaucratic infighting soon emerged to plague the company. Four days after the company took over the model factory site, the Combined Merchant Office of the Furong Saltyard (Furong yanchang changshang luanhe banshichu) petitioned the provincial authority for clarifications. While acknowledging Jiuda's promotion of new cost-cutting technology, the merchants warned of brine shortage driving up costs. Citing the original charter of their model factory approved in 1935 by the Provincial Salt Administration, they demanded that Jiuda's daily output be restricted to no more than 600 *dan*, or some 220,000 *dan* per year.[19] Stepping up their attack, they soon accused Jiuda of taking advantage of the national crisis to encroach on their business, endangering the livelihood of thousands of salt workers. Using steel evaporation pan and coal as fuel, the company was hardly introducing any new technology. A delegation went to Chengdu to demand the return of their factory and found support in Wang Zanxu, the acting provincial governor, as well as other high-ranking Sichuanese officials (and investors in the salt trade).[20] Word soon spread that the protesters were planning to ransack Jiuda's office, and Miao dispatched a squad of armed revenue guards as a precaution.[21] Under pressure, Tang Hansan, the plant manager, recommended relocating the plant.[22]

Fan Xudong characteristically stood firm. He cabled Kong Xiangxi urging him to remove the archaic divisional boundaries so that Jiuda could help increase salt production and gabelle revenue.[23] Seeking to divide the opposition, the company also sought the assistance of the son of Deng Xihou, a former governor of Sichuan (1925–26), in presenting its case to the provincial authorities.[24] Zhong Lijian, who had joined the company, continued to work his magic on Wang Zanxu.[25] As assistant general manager of Jiuda, Li Zhuchen reassured the Sichuan merchants that the company intended to share any innovations and return the plant to them after the war. The provincial military command intervened by sending Xie Guanyi, Tang Hansan's cousin, to investigate and help resolve the dispute. Finally, a bargain was struck: henceforth, existing Zigong producers would enjoy a priority to brine supply, and Sichuan as well as Western Hubei and the six districts of Lizhou in western Hunan would be reserved for Zigong merchants. The company's annual production was reduced to 600,000 *dan* (or 15 percent of the increased production target set by the government) and designated for markets vacated by the Lianghuai division.[26] On the anniversary of the Japanese invasion, the company held an open house for its official launch with five evaporation pans and a daily production capacity of 1,600 *dan*. The choice of date was deliberate and symbolic: the fate of the country and the company was intertwined.[27]

Jiuda's managers, however, were well aware of the difficulties ahead. In his diary, Tang Hansan revealed his worries: compared to Tanggu, Zigong production cost twice as much.[28] Even before it began operations, the company had pleaded with Miao for a steady supply of coal and brine, the former under state rationing and the latter controlled by a local cartel, Furong Rock Salt Well Brine Extraction Co. (Furong yanyajing gongsi).[29] To operate at its designed capacity, Jiuda needed a minimum of 35 tons of coal and 1,400 *dan* of brine each day. Miao Qiujie approved both requests quickly, but the company remained an outcast to the network of Sichuan salt merchants.[30] Despite Miao's support, sporadic supply of coal and brine of low salinity limited the company's production to below 40 percent of its capacity in its first year.[31] According to Tang Hansan, it was a congenital condition that plagued Jiuda throughout the war years. Frequent shortages forced the company to fire only two or three evaporation pans each day, and none at times. However, Fan insisted on fabricating two more evaporation pans to bring the factory up to state-approved production capacity: The government might not fulfill its promises, but he kept his.

Such dedication, however, did not impress officials or ensure their support. Tang Hansan was rebuked for violating protocol: Instead of going through the proper hierarchy, he had briefed the press and cabled Weng Wenhao (and Fan Xudong's neighbor in Chongqing) and Kong Xiangxi directly on the plant's progress. The Finance Ministry, citing Tang's admission in the cable that its production technology might not be entirely novel and its product indistinguishable from ordinary evaporated salt, ordered that the phrase "model" be stripped from the plant's name.[32] The ministry also denied Jiuda's application for an exemption from a charity surcharge on all Zigong salt. The ministry ruled that Jiuda should share the same responsibility for all gabelle and surcharges applicable to other Zigong salt producers.[33]

But not the market assigned to Zigong. The company's first shipment, designated by the state to eastern Hubei, left the plant on October 4, 1938, but by the time it reached Yichang on November 2, 1938, Wuhan had already fallen to the Japanese. The company repeatedly petitioned for permission to divert the shipment to western Hubei and Hunan, reasoning that the government, having approved the establishment of Jiuda in Zigong and exacting the same gabelle and levies, would not deny the company's efforts at a time of war. They were wrong. Sichuan salt merchants lobbied successfully to keep these markets to themselves.[34]

Jiuda was forced to sell elsewhere. Adjusting to the difficult business environment, Fan decided that of the projected twenty *zai* (one *zai* =1,260 *dan*) of monthly production, five *zai* would be sold to the state at the targeted profit margin of one yuan per *dan*. The reminder should be distributed in eastern Hunan directly by the company despite the hazards of wartime transportation and over ten months in turnaround time, or handled by nominee subsidiaries using rented license for open markets in eastern Hubei.³⁵ Through Dingchang, a joint venture with Jincheng Bank, Jiuda began test marketing its salt via Yichang.³⁶

Other attempts at promoting free trade of salt were even less successful. Access to Shaanxi was blocked because it was reserved for saltyards in northern Sichuan.³⁷ The company's attempt to reach Guizhou, then governed by Wu Dingchang, proved equally futile. Wu, Fan's old associate (Fan was also a shareholder of Wu's newspaper *Dagongbao*), and his close aide Zhou Yichun (provincial financial bureau chief and, before the war, vice-minister at the Ministry of Industries as well as a shareholder and a member of the supervising board of Fan's Yongli Chemical Industries) welcomed Jiuda's supply to the provincial capital. Miao Qiujie quickly approved the application, although his order was countermanded, limiting Jiuda's access to those peripheral areas served by the Sichuan–Lianghuai and Sichuan–Guangdong divisions. The setback was traced to Sichuan salt merchants lobbying Xu Kan, a vice-minister of finance who had clashed repeatedly with Miao.³⁸ Caught in the bureaucratic rivalry, Jiuda's shipment was seized until Zhou arranged for the Provincial Cooperative Bureau to distribute the salt for medicinal use.³⁹ Similarly, only a small amount of salt was sold in Yunnan in 1942.⁴⁰ While demand for salt in eastern Hunan remained brisk, long distance transportation during wartime made the trade treacherous, losses high, and lengthy turnaround time.⁴¹

But the company continued to serve the country whenever needed. During the winter of 1940, with the onset of wartime inflation, panic and hoarding began to disrupt the salt market. At the request of the Salt Administration, the company rushed its salt to Chongqing, doing business as Zhongda to mask its involvement. As a precaution against any problems with other merchants, Sun Xuewu appealed for assistance from Lu Zuofu whose brother was then district administrator of Beipei, a model district of Chongqing.⁴² The long lines of anxious citizens disappeared.

Everything for the Country

Frustrated in its effort to increase salt production, Jiuda reversed the business strategy of Zigong merchants. Instead of specializing, it sought vertical integration by diversifying into coal mining and brine well drilling to secure its coal and brine supply.⁴³ Hengfengtang, Jiuda's holding corporation, and Jincheng, its house bank, formed several joint ventures, including for two collieries (at Shawochang and

Gongtanzi, both in Weiyuan) and brine wells: Licheng leased in 1940, Xianzhibei in 1942, Dajiang in 1943, and Hehai in 1944. Marginal in quality and costly to improve, the brine supply from these wells never met Jiuda's needs, with Hehai alone costing 30 million yuan in reboring.[44]

Despite the hostile reception, however, Jiuda kept its promise to share any technological innovation in lowering costs and increasing production.[45] The company introduced electric arc welding to fabricate pans made of steel plate with an evaporation surface as large as 90 sq. m., a design far more efficient than the traditional "two-inch thick cast iron pans weighing over a thousand *jin*."[46] Tang Hansan, working with the group's Golden Sea Chemical Industry Research Institute, attacked the problems of low brine salinity and the high fuel cost for boiling brine. In April 1939, the researchers pioneered use of locally available bamboo to build evaporation shed systems invented in Nauheim, Germany, and first introduced into China around 1910 at Yingcheng, Hubei. Weather vanes were used to determine the optimum configuration to concentrate the brine further to reduce the cost of evaporation by as much as two-thirds.[47]

Other improvements in manufacturing followed. In 1940, Jiuda pioneered drawing brine with electricity generated by an old truck engine. Each pump replaced as many as sixty buffaloes, reducing daily operation cost by almost 30 percent.[48] Further savings came from redesigned furnaces using 20 percent less in coal while increasing production by 30 percent.[49] To lower transportation costs and wastage en route, Sichuan producers had long packaged salt bricks compacted by hand. Researchers at Golden Sea adapted and improved upon the traditional process, replacing the hand-tightened screw-drive press with a motorized press using a wooden piece mold; the salt was dried further by residual heat from the redesigned furnaces. By January 1941, an inexpensive system that cost only twenty-five yuan to build began producing salt bricks of high quality.[50] Over 100 evaporation sheds were in use at Zigong, and plans were made to diffuse the technology to Eastern Sichuan in 1942.[51] True to its word, Jiuda shared all these technologies and impressed visiting foreign dignitaries for its "progressive" work.[52]

Development of the Chemical Industry and Profits

Similarly, the company made good use of its expertise to promote Zigong's chemical industry. Researchers at the Golden Sea Research

Institute traced the cause of mysterious seizures and deaths to barium chloride and found a cheap way to remove the poison from Zigong salt.[53] Systematically, they also analyzed brine samples and the bittern residue from the evaporation process and found a rich brew of magnesium, calcium, potassium, lithium, strontium, rubidium, and cesium which could be processed into chemicals of commercial value: potassium chloride as an ingredient for explosives, boric acid and borax in the manufacturing of glass and disinfectants, as well as bromine and iodine for medicinal use.[54]

Turning industrial waste into marketable chemicals became an integral part of Jiuda's wartime effort. In addition to producing alcohol under contract for the Number 21 Ordnance Factory, Jiuda launched a by-product plant in late 1941, ultimately producing as many as eleven chemicals. In April 1942, a similar facility was established to process bittern from its Licheng brine well. In June 1943, Jiuda, Yongli, and the Golden Sea Research Institute joined forces to launch the Sanyi Chemical Works.[55] Much of the potassium chloride it produced went directly to the No. 23 Ordnance Factory, while borax and boric acid supplied pharmaceutical companies and hospitals.[56]

As word spread that what had been discarded for centuries could be processed into marketable chemicals, however, Zigong's merchants who controlled the brine and bittern supply quickly joined the rush by hiring skilled workers from Jiuda.[57] Without a steady supply of its own, Jiuda and Sanyi had to purchase both brine and bittern to continue their intermittent operation. For 1943, Jiuda's chemical subsidiary contributed a modest 411,087 yuan (or 4.34 percent) to the company's pretax profit of 9,458,747 yuan.[58] Patriotism, at least for Jiuda, did not translate easily into profit.

Indeed, an analysis of Jiuda's production and financial records during the war years reveals a disappointing performance, as shown in Appendix 3a–c. In 1939, its best year during the period, the company produced over 180,000 *dan* of salt, accounting for less than 2 percent of the Chuankang Division production, whereas its seized Tanggu refinery produced 432,437 *dan*, of which 30,520 *dan* was taken by the Japanese military and 244,560 *dan* exported to Japan.[59] Limited by irregular brine and coal supplies, Jiuda in free China operated under one-third of its designed capacity during this period.[60] While the company did turn a profit, it was only a modest one after adjusting for inflation.[61]

The Nationalist's tax policies further eroded the company's profits. Paying the same rate as other Zigong salt merchants but denied access to their designated market, the company paid over 700 million yuan in gabelle and levies from 1939 to 1945 (see Appendix 3b).[62] The government also audited salt production costs as part of its attempt to stabilize wartime prices and thwart profiteering. Although by special approval Jiuda was audited independently to "encourage use of advanced technology," any savings (or potential profits) that accrued were eliminated by a cost-reduction levy (*chengben jianqing fei*) promulgated to prevent wartime profiteering.[63]

Relentless Japanese pressure complicated the salt business even further. With the fall of state-mandated transportation routes to the Japanese or local bandits, the company found reaching the market in eastern Hubei and Hunan increasingly difficult, forcing it to sell more and more to the state directly.[64] Eschewing black marketeering and multiple ledgers for tax evasion, Jiuda struggled to make ends meet as wartime inflation raged.[65]

Nationalization

The war also presented the Nationalist government with the opportunity to nationalize the salt industry. When war broke out, the merchants could neither pay the requisite gabelle for the immediate release of all the salt stockpiled in the government depots on the coast, nor organize the emergency evacuation of this strategic material.[66] The task of removal and transportation fell on the government. What had begun as an emergency measure became the first step in the nationalization of the trade to stabilize prices and to prevent shortages.[67]

Soon after the war broke out, the Salt Administration began planning for the nationalization of salt production, transportation, wholesaling, and retailing.[68] The need to ensure people's livelihood justified the centralized control of this and other strategic materials. In 1939, the Nationalist Party Central Committee plenary meeting approved the policy on the promise of adjusting "production and consumption of salt as the needs of national defense warrant," stabilize salt prices, rationalize distribution, minimize merchants' windfall profits, and simplify revenue collection.[69] To finance production, purchase, and transportation of 20 million *dan* of salt, the plenary committee budgeted a state capitalization of 3 billion yuan (although 1.5 billion was actually required, assuming a turnover rate of twice a year and zero inflation). The annual gabelle yield of 80 million would be incorporated into the state-set selling price plus a projected surplus/profit of 300 million.[70]

The plan was met with skepticism, if not opposition, from what remained of the foreign staff at the Salt Administration. With war raging, O. C. Lockhart found it a "peculiarly inopportune time" to undertake such a major reform for the purpose of raising "funds to cover a part of the Treasury deficits." He warned that the state takeover would disrupt the existing salt trade, put salt merchants and their capital out of work, create a vastly expanded salt bureaucracy, which the depleted foreign staff could not supervise, and fuel inflation. Under the plan, salt revenue became "a variable commercial profit" rather than a tax, creating a legal problem for creditors who held Chinese government bonds secured on the gabelle. He warned that such a violation of the terms of the Reorganization Loan would "affect unfavorably the sympathetic attitude of lending interests toward China's financial difficulties..." even after the return of peace. Instead, he proposed a general gabelle increase of five yuan per *dan*.[71] After his resignation—he and Zhu Tingqi had been at "complete loggerheads" for years—R. D. Wolcott continued the campaign as acting associate director.[72] Writing to Manuel Cox at the Currency Stabilization Board, Wolcott complained that the Nationalist government was depriving the salt merchants of their "age-old rights." Even worse, he predicted that government borrowing would double under the plan and contribute to a vicious circle of inflation with "grave consequences to the national currency and to the livelihood of the people."[73]

To the Nationalists, however, the timing was perfect. With Lianghuai, Changlu, Shandong, and Zhejiang divisions occupied and the salt trade taken over by Japanese military and collaborators, the revenue farmers had outlived their usefulness.[74] Their hereditary privilege could now be revoked without any compensation and their monopolist profit absorbed as state revenue. Supervision and management of salt production, collection, transportation, and distribution also provided jobs for a fair number of displaced bureaucrats. Steered by Western-trained technocrats with a "mental model" of a controlled economy and state enterprise, the ministry brushed aside the advice of its foreign employees.[75]

On January 2, 1942, the "age-old rights" of revenue farmers came to an end. From 1942 until 1945, the Salt Administration managed the bulk of the trade in Free China, as presented in Table 7.1. During the same period, gabelle, profit, levies, and surcharges from the new system generated over 80 billion yuan in revenue, although after adjusting for inflation, only a fraction (approximately five percent) of its 1937 purchasing power. Nevertheless, as shown in Table 7.2, during the waning years of the war salt accounted for almost half of the government's tax revenue and provided the bulk of the Nationalist government's income from state-ran monopolies.[76]

Under the nationalization scheme, Jiuda became a contract supplier to the state: three *zai* for Chongqing and another three *zai* for Hunan each month.[77] For 1943, the state advanced Jiuda 44.736 million yuan for its salt, leaving a thin gross profit of 5.642 million after deducting 39.095 million in costs, or a nominal profit margin of 12.6 percent. However, by the time the state completed its audit of the company's costs, the purchase price set and final payment made, rapid inflation had eroded the "reasonable" profit margin allowed.[78] To make ends meet, the company also served as a transport tender (*daiyun*) of state-owned salt, receiving reimbursement for insurance and cost of transportation plus interest and a wastage allowance.[79] Jiuda thus

Table 7.1 State salt collection and distribution, 1942–1944 (unit = *dan*)

Year	Production	Collection	Distribution
1942	21,768,225	8,853,274	20,201,925
1943	25,080,168	16,126,538	22,270,306
1944	16,605,268	15,250,776	16,035,143

Table 7.2 Salt-derived revenue for the Nationalist government, 1937–1945[a] (unit = million yuan/*fabi*)

Year	Salt-related	Total Tax Revenue	Salt as % of Total
1937	218	728	29.89
1938	139	469	29.54
1939	113	525	21.58
1940	105	739	14.22
1941	125	996	12.59
1942	1,437	4,475	32.12
1943	1,645[b]	12,169	24.8
1944	17,447[b]	39,180	44.53
1945	61,909[b]	117,059	52.89

Source: [a]Ding Changqing, et. al.(1990), 407. Including army subsidies and other surcharges, Dong Zhenping (2004) retabulated the percentages as 42 for 1942; 47.1 for 1943; and 53.5 for 1945, respectively.
[b] "Monopoly profits" were listed as 1,668,972,000 for 1943, and 1,787,293,000 for 1944. See Guojia shuiwu zongju (1999), 225.

survived on government advances and bank loans amounting to over 90 million at interest rates as high as 84 percent per month by 1944.[80] The state and companies connected to public and private banks came to control the industry through their access to cheap credit and the black market.

State and Private Enterprise

But if Fan Xudong, as a patriotic capitalist, was willing to forego profits, nationalization affected Fan and his company in other ways. Since its inception, Jiuda's operation was vertically integrated—from its own salt flats to manufacturing, distributing, and marketing. When the company applied for registration renewal in 1943, the Finance Ministry belatedly found it in violation of existing regulations by combining production, transport, and distribution.[81] To comply with the law, the company divided itself, legally and for accounting purposes, into the Jiuda Zhiyanchang [Jiuda Salt Refinery] for production, and the Jiuda yanye gongsi [Jiuda Salt Industries] as a distributor, transporter, and operator of official retail outlets.

Accepting the subsidy from the Nationalist government to evacuate the group, too, carried unforeseeable risks. While in Hong Kong negotiating for a license of the Zahn process to produce ammonia sulfate (as an explosive and fertilizer) and soda ash in Sichuan, the Ministry of Finance surprised Fan with a notice on April 23, 1938: the remaining 2.6 million yuan could be released en bloc to Fan, subject to the conversion of the entire 3 million yuan subsidy into shares, creating a joint public–private enterprise, then a novel form of limited liability corporation with the state as the majority shareholder.[82]

The politics of this conversion, or who was behind the attempt, remained opaque.[83] However, Fan Xudong would not tolerate this state intrusion, and he cabled Kong Xiangxi declining the offer. Capitalization of his companies had been paid-up in full, he argued, and any increase required a shareholders' meeting. Given the raging war and shareholders stranded in occupied territory, such a meeting could not be called, at least for the time being. He also appealed to Weng Wenhao who reported two days later of his futile efforts to intercede.[84] Citing a conflict of interest, Fan replied the same day to decline the subsidy (or state capital infusion). He would rather devote his energy to rebuilding his group. Weng then cabled on May 13, 1938, requesting a face-to-face meeting to resolve the impasse, and Fan flew back to Hankou two days later.

Existing archival sources are silent on subsequent negotiations, but Fan reported four days later that the government had approved his plan to consider the infusion of state capital at a shareholders' meeting to be convened as soon as feasible. After another round of lobbying, the subsidy was finally released by installment beginning June 13, 1938.[85] For the time being, then, Fan Xudong averted immediate annexation by the Nationalist state, a fate that befell Liu Hongsheng and other entrepreneurs.[86]

But there was a limit on how much Fan could bank on his network and his patriotism. On September 26, 1943, he met with Jiang Jieshi and hand-delivered his most ambitious plan yet.[87] Adding his voice to postwar reconstruction, no time should be lost in preparing a national plan and training a new generation of engineers. He offered his expertise to build ammonium sulfate, coke, and soda ash plants, ten in all, scattered throughout the country, including Zhuzhou, Hunan; Yanxing, Yunnan; Tongguan, Shaanxi (or Lanzhou, Gansu); and Jingxing, Hebei. When completed, he expected the country to be self-sufficient in the basic chemicals crucial for agriculture, national defense, and industrialization. All he needed was a $10 million credit line through the Universal Trading Corp., and a 15 million yuan bank loan guaranteed by the Finance Ministry.[88] His proposal was motivated purely by service to the country—ownership and operation of the plants were matters for the state to decide.[89] A skeptical Chen Bulei, Jiang's confidential secretary, found the whole proposal unreasonable (*buheli*). The nature and significance of these plants to national defense made private control undesirable, and he questioned why, if the state financed the projects, the Ministry of Economics should not assume direct control. Chen recommended that Weng Wenhao should deflect Fan. Jiang concurred and directed Fan to discuss the proposal with Weng and Kong Xiangxi as well.[90]

Undaunted by this bureaucratic runaround, if not quite rebuff, Fan wrote to Weng on October 11, 1943, requesting a meeting under the impression that he had Jiang's approval "in principle."[91] By then, Weng had leaked the proposal to Qian Changzhao, his deputy at the National Resources Commission, who filed a scathing confidential memo with Jiang on October 15, 1943. Ignoring the wartime difficulties, harsh words littered Qian's memo: Fan enjoyed an undeserved reputation (*xuming*) as a patriotic entrepreneur. The plants in Tanggu and Nanjing had not been destroyed by the Japanese, but Fan received 3 million yuan as a subsidy for his loss (or share from the government's perspective) that Qian had helped secure. Fan had also promised to construct new plants at Wutongqiao, all unfulfilled to date except for the salt refinery. Given this record, further support of Fan would discourage "real" entrepreneurs. Ultimately, Qian, a graduate of the London School of Economics, argued that these basic chemical industries, vital to national defense, must not be in private hands.[92]

Undeterred, Fan Xudong soldiered on. In August 1944, Jiang Jieshi chose him as one of the delegates representing China at the International Business Conference. Before their departure, Chen Guangfu, Lu Zuofu, Bei Zuyi (replacing Zhang Gongquan who became an adviser), and Fan (the fifth delegate, Li Ming, was already in the United States) met with Jiang to ascertain the government's position toward the main issue on the conference agenda: how to preserve and promote private enterprise, especially in developing countries.[93] Jiang promised

> ...to exercise a liberal policy...projects of transportation, hydraulics...big power houses and sea harbors...will have to be taken care of by the government. Within ten years...China will open her doors to foreign investors...and it will not be necessary for sino-foreign concerns to be required to have Chinese chairmen on the board with Chinese presidents.[94]

In other words, foreign investment would be welcomed.

Expounding on Jiang's position, the delegates declared at the conference that they were

> strongly in favor of free enterprise and free trade, but believe that in the transitional period from war to peace, a certain amount of control will be necessary to conserve our limited resources of foreign exchange. Some industries and projects that have a direct bearing on national defense and those that are beyond the financial resources of private business will be owned and controlled by the government.[95]

Now that he could not be criticized for a change of heart in courting foreign investment, Fan Xudong also redefined his patriotic capitalism to accommodate another necessary evil. After the conference, he approached the U.S. Export-Import Bank in Washington which had earmarked $500 million for loans to China.[96] On January 22, 1945, Fan cabled Jiang directly about his negotiations for a $16 million line of credit, secured by the equipment purchased, at an annual interest rate of 4 percent, and repayable over 15 years.[97] Both Weng Wenhao and Chen Bulei considered those terms favorable and recommended to Jiang that Fan be permitted to proceed, provided that the government was informed of his progress and the details of the contract.[98]

Old China hands at the State Department, led by O. C. Lockhart, remained skeptical, however. In terms of war material, chemicals ranked low in priority. They argued that Fan's group had accepted capital from the government, and the Eximbank should consider "the question of government versus private investment in the company." "Extension *at this time* (emphasis in original) of credits to finance exports to China was ill-advised."[99] Fan managed to convince Warren Lee Pearson, president of the bank, that his group was privately held. The technicalities raised by the State Department were "properly matters for decision by the bank."[100]

Fulfillment of Fan's lifelong quest to make China's chemical industry independent seemingly was at hand. An impressed Chen Guangfu captured the entrepreneur's buoyant spirit:

> Fan is very happy about the successful negotiation…and his trip…can be considered a fruitful one. He has always felt that he has been throttled in his undertakings during the past years due to (a) lack of capital. His thirty years' effort and hard struggles have enabled him to produce good work of Yongli which is directly responsible for the contract of the loan. He has at last reaped the harvest for his loyalty to his projects.[101]

Upon returning to China, Fan planned to report to Jiang Jieshi, build his plants, invite Chen to join the company's board, and retire the loans by exporting to Southeast Asia. He was confident of the state bank guarantee, because, without his plants, China would have had to pay for chemical imports anyway.

On May 1, 1945, Fan Xudong signed a contract with the Eximbank. Two days later, Fan cabled Weng Wenhao a summary of the contract. Clause 4 specified that repayment of principal and interest would begin in 1949 over ten semiannual installments "to be guaranteed by a state bank designated by the Chinese government."[102] On May 22, 1945, Weng forwarded the cable without comment.[103]

Weng Wenhao's filing set off alarm bells at the generalissimo's office. Neither Fan nor Weng's prior communications referred to a state guarantee, and the ministry was ordered to obtain and review the contract.¹⁰⁴ By then, Fan had already left New York with Hou Debang, stopping in England for a friendly exchange of views with Imperial Chemical Industries, then on to India where Yongli held a consulting contract with Tata Industries.¹⁰⁵

Existing archival sources are silent on why Fan did not alert Chongqing in early April when the main terms of the contract, including the state guarantee clause, were already settled.¹⁰⁶ Perhaps he had lost touch with politics in Chongqing, and/or he was overly confident of his cause. Whatever his reasons, when he returned to Chongqing on June 22, 1945, and a copy of the contract finally filed a week later, Fan's triumph turned into frustration. Fan found Weng Wenhao, his friend and neighbor in Chongqing, a changed man.¹⁰⁷ Song Ziwen, appointed June 1, 1945, as head of the Executive Yuan (with Weng as deputy while concurrently minister of economics), had grand designs for a $2 billion aid package under the direction of the National Resources Commission. In August 1945, Song visited Washington, D.C. with Qian Changzhao to negotiate the package.¹⁰⁸ They, too, encountered many skeptics at the State and Commerce Departments.¹⁰⁹ American businesses, while welcoming the investment opportunities, were concerned about China's "illiberal" policy toward foreign investors and the prospect of competing with state-controlled enterprises financed by U.S. government loans.¹¹⁰ Song left Washington frustrated, and so should Fan.¹¹¹ The Executive Yuan directed the Ministry of Economics to develop a comprehensive policy on state guarantees of foreign loans to private Chinese enterprise, and promptly tabled the draft.¹¹²

As the debate over the relative merits of state control and private enterprise flared again in China and abroad, Fan's bank network, too, was crumbling. The war had already cost him his friendship with Zhou Zuomin.¹¹³ With the end of war in sight, Zhang Gongquan, Li Ming, and Chen Guangfu again joined forces to form China Industries, Inc. with General Electric, Lazard Frères, and Lehman Brothers. Separately, Chen Guangfu was negotiating with the Pennsylvania Salt Manufacturing Co. Ltd., and the Imperial Chemical Industries for joint ventures in China, including his own caustic soda plant.¹¹⁴ Caught in the crossfire between Washington and Chongqing, once supportive officials such as Weng Wenhao and Qian Changzhao turning against him, and his bank network's loyalty divided, Fan's request languished in the Nationalist bureaucracy amidst celebrations of the Japanese surrender, his last plea to the Ministry of Economics dated August 31, 1945. He died after a brief illness on October 4, 1945.¹¹⁵

Fan thus remained true to his patriotic capitalism and the third path. During the war against Japan, he could have stayed and maintained control of his companies through Chinese or Japanese middlemen, or both, made money, and even expanded, as Song Feiqing of Dongya fame did; alternatively, he might have continued under British cover (at least for the time being), or managed them through an adroit Zhou Zoumin.[116] Instead, he chose to evacuate and dedicate his most important assets—the managers, engineers, and researchers—to a new mission of industrializing the Southwest in support of the war effort. However porous the boundary between resistance and collaboration, Fan did not cross that line.

But patriotism exacted its toll. Risking state annexation, Fan took subsidies and loans from the Nationalist government to ensure the group's survival amidst growing hostility. True to its words, Jiuda shared its scientific and technological innovations for the common good, but Sichuan merchants guarded their turf jealously in the competition for brine, coal, markets, and profits. As war raged, the needs for revenue, stable prices, and continued supply led to the nationalization of the salt industry. To Fan Xudong, whose crusade against hereditary salt revenue farmers and government-controlled industrial enterprise had occupied much of his adult life, the victory must have been bittersweet, if not hollow. With the demise of the revenue farming system, he won his life-long war.[117] On the other hand, free marketing of salt and development of the country's chemical industries, his other dreams, remained unfulfilled.

Indeed, when peace returned, Jiuda was a wounded company confronting a changed business. Losses during the Second World War amounted to millions of yuan.[118] Following Fan's death, Hu Shi was elected chairman of Jiuda, and Fan Hongtao, Xudong's brother, served as acting general manager. Although the Nationalists declared the free trade of salt and reiterated the abolition of revenue farming in 1945, the salt trade had come under the control of companies operating as subsidiaries of big banks, both private and state-owned, such as Daye (Chen Guangfu's Shanghai Commercial and Savings Bank), Dachangyu (Bank of Communications), Zhonghe (Farmer's Bank), Dayou (Shanxi Yuhua Bank/Kong Xiangxi), or as investments by various branches of the Nationalist party (Yongye, on behalf of the Central Committee; Anli, on behalf of the Shanghai branch). The biggest of them all, the China Salt Industries Corp., Ltd. (Zhongguo yanye gongsi) boasted of 1 trillion yuan in capital and, with all the salt-related assets seized from the Japanese, controlled 75 percent of the country's production capacity. As a nominally joint state–private enterprise headquartered in Tianjin, its board was dominated by officials representing state shares, including Miao Qiujie, Zeng Yangfeng, and Yao Yuanlun (three of the four "guardian gods" of the Salt Administration), and former officials such as Ma Taijun (as general manager of Kong Xiangxi's Dayou). Representing "private" investors were the big banks, both state and private, such as Bank of Communications (Qing Yongming), Shanghai Commercial and Savings (Li Tongcun), Bank of China (Xu Weiming), Farmer's Bank (Li Shuming), and Central Trust of China (Wu Rencang), as well as the politically influential, including Li Mingyang (Sino-American Industrial Development Co., Ltd. on behalf of Song Meiling), Chen Qicai (uncle of Chen Lifu of C.C. clique fame), and Wang Tianxing (Nationalist Central Committee).[119] Fan Hongtao, as a charter shareholder with a 200 million yuan stake, did not even make the board.[120] Caught between official nationalism and powerful banking interests, Jiuda languished (see Appendix 4a on the company's performance during these years).

But the worst was yet to come. Seven months after the Ministry of Economics ordered restitution of the company to its rightful owner, Miao Qiujie, the company's nemesis before the war and hospitable host during the war years, filed a confidential memo with the Executive Yuan seeking to modify the ministry's decision.[121] Ignoring the fact that the Japanese had seized Jiuda at gunpoint and the damages it sustained during the war years, Miao, acting director of the Salt Administration, argued that the company was an "enemy enterprise." On April 1, 1943, Jiuda (Tanggu) had been released from military occupation and became a partnership between the collaboration regime (Huabei zhengwu weiyuanhui) and Kahoku engyō kabushiki kaisha (a subsidiary of Kita Shina haihatsu kabushiki kaisha), a move that Jiuda in Chongqing had declared at the time as illegal.[122] With a Japanese-dominated board and majority stake, the Japanese made a "massive" capital infusion to finance Jiuda's production for exports to Japan and military needs.[123] As the Salt Administration under the Finance Ministry was to take over the Kahoku engyō kabushiki kaisha and all its assets, Miao argued that the decision of the Ministry of Economics be vacated, and that portion of Jiuda's assets attributable to Japanese investment be converted into company shares, making the state its majority shareholder.

Tapping into Fan's network, Li Zhuchen and Hou Debang made personal appeals to Zhang Qun, then head of the Executive Yuan, for his good offices in preserving Jiuda as a private enterprise.[124] The company had suspended transfer of its shares since the war began. There was no question of any Japanese shareholders or ownership.[125] Strategically placed friends and secretaries in various ministries also offered advice and leaked documents.[126] The company presented as evidence the ruins of its factories and scoured its ledgers to highlight wartime losses, rendering the debate over "subsidy" or "share" somewhat irrelevant. Finally, the 3 million yuan "subsidy" was repaid in full with 6 billion yuan on April 9, 1948, facilitated by a separate 2 billion yuan donation to the Young China Party whose leader, Chen Qitian, then served as minister of commerce and industry.[127] Jiuda remained a private company as Fan Xudong wanted it.

Postscript

Yet memory of Fan lives on. Jiang Jieshi's eulogy "Earnest and dedicated in practice" (*lixing zhiyong*) graced his memorial service, and Mao Zedong proclaimed him an "industrial pioneer to China's credit (*huagong xiandao gongzai zhonghua*)."[1] Such was Fan Xudong's selfless dedication to the common good that Hu Shi anointed him a modern day sage (*xin shengxian*) equal to any ancient ones.[2] Amidst foreign imperialism, Japanese invasion, and political chaos beyond the control of any of the actors in the period, Fan kept working: his enterprises were, in his own words, a never-ending struggle.[3] He was also remembered as a patriotic capitalist whose premature death was caused, directly or indirectly, by a callous regime.[4]

Indeed, charting a third course of national development between state and private capitalism, Fan and his colleagues conceived and nurtured Jiuda as an instrument of national economic development and socially responsible capitalism. Not a mere profit-maximizing private enterprise, the company served several missions: competing against a tax-free foreign import, offering a modern and hygienic alternative to consumers, promoting reform of the gabelle system, and, as a "Chinese elder brother," financed research and development of the country's chemical industry, all for what Fan hoped would bring national salvation.[5] Confronting the many obstacles in its path, the company's microhistory reveals several intertwined processes: the embeddedness of Fan and the company in culture and socioeconomic institutions; management of market, hierarchy, and network as a part of entrepreneurship; and the complex relationship between private enterprise and the state in modern China.

Using refined salt as a catalyst for change ran afoul of both the gabelle administration and the revenue farming system. In addition to opposition from deep-pocketed revenue farmers, Fan and his colleagues had to contend with the Salt Administration, the Revenue

Inspectorate, and various provincial governments and political chaos. Despite its modernizing mission, the synarchic international bureaucracy, charged with raising revenue to satisfy the foreign bondholders of the Reorganization Loan and generating surplus to maintain its hold over the republican governments, often found itself at odds with Jiuda's goals of promoting a scientific, hygienic, and modern product. Even after 1949, Beijing found divisional boundaries and state monopoly of the salt trade not incompatible with its revolutionary goal of economic planning. Indeed, revenue targets combined with state-run salt companies and financing remained the modus operandi of state enterprise for years to come.[6]

Promoting institutional change thus required more than patriotism and the rhetoric of modernity, or for that matter, market forces of supply and demand. Fan and his colleagues managed their portfolio of socially embedded ties, both strong and weak, to mobilize resources and solve problems. When these proved inadequate, relationships were manufactured through "pay to play," gifts, and ultimately, bribery. Even after 1949, Li Zhuchen drew on his insider status as a high-ranking member of the new regime to keep Jiuda afloat. Yet, all these could be costly and the outcome uncertain. Limited in reach by a sharp drop-off, the transitive law of networking also exposed the company to political risks, shuttering it several times, and the kidnapping of Fan. Networking both enables and constrains.

To the extent Fan and his colleagues managed their network effectively, they forged a hybrid business organization combining elements both traditional and modern, Chinese and Western. Registered as a limited liability company, Jiuda also operated Hengfengtang, an unlimited liability holding corporation, to evade taxes and regulations, and invest in land, salt pans, and partnerships with revenue farmers and local franchisees. While mobilizing its shareholder network for a variety of resources, the company held annual shareholders meetings, elected board members and supervisory committees, and discussed novel practices such as stock splits and bond offering. Fan, like his contemporary such as Liu Hongsheng, also consulted with family and subordinates, codified regulations, and delegated authority. Successive management reforms introduced a hierarchical structure and Western accounting, although networks based on kinship, native place, and school remained strong in the governance and operation of the company. While using labor bosses to manage its temporary workers, Jiuda was also among the first in the country to introduce an eight-hour work day and offered a host of benefits to its staff and workers, including housing, literacy classes, recreational facilities, health care, and tuition-free schooling to forge the company into a socialized unit. Adding to the debate over theories of simple convergence or divergence, Fan thus forged a crossvergent, perhaps even a transvergent, business organization characterized by "transformative reinterpretation and application of the indigenous cultural perspective."[7]

Fan also confounded his contemporaries, both foreign and Chinese, as a "difficult to deal with" entrepreneur. He stomached no lavish meals, nor entertained his network with elaborate banquets. Soured by his brief tenure at the National Mint and negotiations with officials, Fan repeatedly declared that it was "inadvisable for the government to participate financially in or in any way control any industrial scheme," much less any involvement of bureaucrats in his enterprises.[8] Unlike the Nationalist party and other advocates of China's industrialization through state enterprise, Fan harbored no such illusions.[9]

Ever short in capital, he lobbied for state subsidies and guarantees of private bank loans, and during the Second World War, even state loans, but government involvement in his enterprises was out of the question. Jiang Jieshi's dossier on Fan read, in part,

> Determined and forthright in his views...Among the handful steeped in the theory and practice of economics...Neutral in politics. Although critical of planned economy and other economic policies, (he) supported the central government.[10]

Yet, Fan also declined ministerial appointments twice while serving as a member of various state commissions, committees, and the People's Political Council.[11] He walked a fine line between serving the country and the state.

The prosopography of Fan and his colleagues—Li Zhuchen, Sun Xuewu, Hao Debang, and others—also tells the collective story of a generation of entrepreneurs educated abroad. Wealth and power beckoned them upon their return to the country, yet they led austere lives, with no houses or automobiles they could call their own. Following the example set by Fan, they declined official positions, substantial salaries, and bonuses. Li turned down Standard Oil to work at Jiuda. Sun left the English-controlled Kailan Mining Administration, took a pay cut, and declined offers from Song Ziwen, his roommate at St. John's University and Harvard. Along the same vein, Hou took not one cent of the $25,000 consultant's fee from Tata Industries. Drawing on Jiuda and other resources, they dedicated themselves toward the industrial development of the country. At least 1.3 million yuan of Jiuda's capital was diverted illegally, according to the 1929 Company Code, to Yongli. When ICI and IG Farben Industrie AG demanded a hefty design fee, a royalty, tax exemptions, guarantees, and monopoly privileges to build an ammonium sulfate plant in China, Fan and his colleagues again met the country's need by raising the necessary capital from private banks. Yongli would assume all the risks except for a Nationalist government guarantee of any shortfall in interest payment for three years.[12] Patriotism itself could be a strong motivation for economic development.

Fan's blending of patriotism and capitalism thus made a peculiar combination. If profit maximization and self-interest were the only guiding principles to economic behavior, and the capitalist intentionality did not matter, Fan's version required that, for the benefit of the country, no sacrifice was too great. His many industrial schemes were designed to strengthen the country, to fulfill one's duty as a Chinese, and not for profit, much less personal gain.[13] Under Fan and Li Zhuchen, Jiuda became an icon of modern China's "national bourgeoisie" enterprise. On the other hand, their attempt to distinguish the country from the state did not endear them to either the Nationalist or the Communist state.

Last but not least, Jiuda's experience also offers a comment on the controversies surrounding Chineseness and business in modern China. Garden varieties of capitalism have blossomed beyond England and America since the eighteenth century, while the recent success of the Chinese economy has added another bloom.[14] Emphasis on education, family, social harmony, Confucian work ethics, ethnicity-based network, and hierarchical (if not paternalistic) management are declared by some scholars as the defining characteristics of Chinese business, just as other scholars emphasized bribery, network, and cronyism "built on systemic corruption and raw political power."[15] On the other hand, other scholars found that China's situation is not "really outstanding in comparison to other world regions" and could have its place in an imperfect market to promote economic efficiency and development.[16]

Indeed, as a critique of cultural essentialism and orientalism (imposed or self-orientalizing, or both), scholars have questioned whether many defining characteristics of Chinese capitalism are uniquely Chinese.[17] Others have gone even further, arguing that culture and history of institutional arrangement did not matter, or at least make little or no difference in global capitalism. Globalization and the resulting grobalization are nothing but capitalism practiced by persons of Chinese descent.[18] On the other hand, as globalization unfolds, the tension between convergence and glocalization might lead to crossvergence of business practices, whether as a transitional stage or a more lasting formation, perhaps even transvergence. To Fan Xudong and his colleagues, it was a selfless public mindedness that defined Jiuda as a Chinese business.

Appendix 1

Capital, reserves, and profits of Jiuda, 1916–1937[a]

(Unit: yuan)

Year	Paid-up Capital[b] (c)	Reserves	Net Profit (p)	Nominal Profit Rate p/c %
1916	66,295	na	na	na
1917	150,000	24,580	23,086	15.4
1918	300,000	66,762	75,024	25
1919	500,000	101,762	137,593	34.5
1920	500,000	225,428	171,959	34.4
1921	1,700,000		308,165	18.1
1922	1,700,000	46,368	426,191	25.1
1923	1,700,000	89,959	430,050	25.3
1924	2,100,000	136,060	424,518	20.2
1925	2,100,000	176,579	361,206	17.2
1926	2,100,000	219,425	345,072	16.4
1927	2,100,000	250,998	278,289	13.3
1928	2,100,000	277,800	283,096	13.5
1929	2,100,000	307,997	122,544	5.8
1930	2,100,000	224,730	141,831	6.8
1931	2,100,000	154,412	152,060	7.2
1932	2,100,000	92,873	212,285	10.1
1933	2,100,000	73,475	301,924	14.4
1934	2,100,000	116,823	286,289	13.6
1935	2,100,000	149,228	310,164	14.8
1936	2,100,000	198,345	316,893	15.1
1937	2,100,000	192,757[c]		

Sources: [a] Company reports held at the Tianjin Jianchang archives (Tanggu), and Archives of the Jincheng Bank held at TM and SM.
[b] Paid-up capital of Jiuda, including investments in Dapu, Yongyu and Yongli at book value, although the amount diverted illegally to Yongli remains an accounting mystery. Of the 1.2 million yuan capital increase of 1920–21, the "bulk" went to Yongli, as did the 400,000 raised between 1923–24.
[c] January to July 1937.

Appendix 1

Capital, reserves, and profits of Jugo, 1916–1927

Appendix 2

Profits and dividends of Jiuda 1917–1936[a]

(Unit: yuan)

Year	Net Profit Shipped dan p/dan	Dividend Rate[b] (as % of 100 yuan face value @share)
1916	na	
1917	1.869	8
1918	0.944	30
1919	0.877	24
1920	1.160[c]	24
1921	0.999	12
1922	1.040	na
1923	1.021	15
1924	0.964	14
1925	0.981	10
1926	1.192	10
1927	0.819	8
1928	0.865	8
1929	0.430	8
1930	0.618	8
1931	0.577	8
1932	0.633	8
1933	0.617	8
1934	0.604	8
1935	0.378[d]	8
1936	0.502	8

Sources: [a] Company reports to the Changlu Salt Division held at the Hebei Provincial Archives, and Archives of the Jincheng Bank held at *TM* and *SM*.
[b] *Sources*: Company papers, and Archives of the Jincheng Bank held at *TM*. Original company charter specified a dividend rate of 7 percent.
[c] Revised from 0.709 yuan per *dan* because of a special transfer of 100,000 yuan to reserves.
[d] Shipment doubled pursuant to government requirement of ever normal salt reserve.

Appendix 2

Profits and dividends of Jute 1917–1938

Appendix 3a

Production, Jiuda Salt Industries (Zigong) and Chuangkang Division, 1938–1945[a]

(Units:1 *zai* = 1,260 *dan*)

Year	Production *zai/ dan*	Chuankang[b] *dan*
1938 (Sept.–Dec.)	33 / 43,315	8,463,000
1939	143 /180,545	9,377,000
1940	101 /128,003	9,735,000
1941	/120,000[c]	9,878,000
1942	/108,420[c]	9,302,000
1943	81 / 94,827	8,682,000
1944	102 /	8,558,000
1945 (Jan.–July)	34.5	8,462,000

[a] *JD* 100908–100914, 300601, 300645, 300650, 300651, 300657, 300663.
[b] *ZGJDYWSZL* IV: Table 2, 241–2.
[c] (Tang) Hansan, "Zaijing liunian," [Six Years at Zigong] *Haiwang* 16.31 (1944), 246.

Appendix 3b

Jiuda Salt Industries (Zigong): Taxes, levies, net profits 1938–1945

(unit = yuan)

Year	Gross Salt Sales	Reduced Cost Levy	Taxes and Other Levies	Net Profit[a]
1938	133,627	9,269	N.A.	N.A.
1939	3,303,821	15,069	2,285,460	120,369
1940	4,766,467	153,650	2,658,481	194,970
1941	15,751,480		8,398,448	669,183
1942	40,941,835		15,359,914	412,282
1943	67,963,513		27,057,022	7,413,436
1944	334,559,072		194,507,399	7,217,971
1945	784,386,916		512,908,851	31,622,728

[a] After tax and depreciation, including its salt, chemical, engineering, and transport subsidiaries.

Appendix 3c

Jiuda Salt Industries (Zigong): Net profits adjusted for inflation 1938–1945

(unit = yuan)

Year	Inflation Index[a]	Net Profit (constant 1938)
1937 (Jan.–July)	192,757	
1938	1	N.A.
1939	1.29	93,309
1940	4.68	41,660
1941	19.37	34,547
1942	44.40	9,285
1943	125.87	58,897
1944	436.54	16,534
1945	1,308.07	24,175

[a] Inflation index, with 1938=1. Adopted from Appendix Table 2.4, "Wholesale price index of Chungking" in Shun-hsin Chou, *The Chinese Inflation, 1937–1949* (New York: Columbia University Press, 1969), 306.

Appendix 4a

Jiuda Salt Industries (Shanghai): Net profits adjusted for inflation 1946–1948

(unit = *dan*/yuan/golden yuan*)

Year	Salt Produced (dan)		Shipped	Net Profit
	Refined	*Washed*		
1946	162,248	128,142	210,677	329,384,269
1947	307,233	417,236	737,787	31,612,000,000
1948	175,496	510,640	836,684	1,885,993*

Appendix 4b

Jiuda Salt Industries (Tianjin): Salt production/Net Profits 1949–1954

(unit = *dan*/old *renminbi*)

Year	Salt Produced (dan)			Shipped	Net Profit (Loss)
	Refined	Washed	Raw		
1949	36,090	136,102		547,056	2,431,258,234
1950	76,030	241,711	518,418		56,273,484[a]
1951					(2,900,688,989)
1952					(1,420,450,261)
1953					5,018,320,603
1954					10,375,905,142

[a] With inventory marked-to-market at 3,070,492,625 See JD 300198.

Appendix 4c

Jiuda Salt Industries (Tianjin): Balance sheet, April 31, 1950[a]

(unit = old *renminbi*)

Debit		Credit	
Bank Loan	1,450,000,000	Raw salt	3,165,877,305
Gabelle due	7,694,730,000	Refined salt	510,796,800
Misc. debts	1,375,000,000	Washed salt	808,002,000
Accounts payable	3,948,325,500	Spring salt	
		Harvest	2,200,000,000
		Inventory	7,030,155,460
		By-products	1,217,144,200
		Cash/deposit	95,819,658
Subtotal	**14,468,055,500**		**14,801,795,423**

[a] JD 200034.

Appendix 4d

Jiuda Salt Industries (Tianjin): Balance sheet, September 20, 1950[a]

(unit = old *renminbi*)

Debit		Credit	
Bank loan	7,200,000,000	Raw salt	4,000,000,000
Gabelle due	3,873,760,000	Refined	1,100,000,000
Misc. debts	714,414,000	Supplies	2,000,000,000
Accounts payable	3,466,834,700	Inventory	8,473,418,500
		Autumn salt	
		Harvest	980,000,000
		Cash/deposit	539,653,375
Subtotal	15,258,008,700		17,093,071,875
		Fixed/assets	
		Equipment/land	18,000,000,000
		Yongli shares	150,000,000,000
		Yongyu shares	15,000,000,000
		Victory bonds	2,715,000,000
		Subtotal	185,715,000,000

[a] JD 300325-6.

Appendix 4e

Jiuda Salt Industries (Tianjin): Assets and liabilities, 1953[a]

(unit = old *renminbi*)	
Fixed assets (net)	14,741,479,000
Fixed value assets	4,762,933,083
Other	86,720,757,678
Total assets	106,225,169,761
Liability	9,730,751,887
Net assets	96,494,417,874
Japanese/puppet government	7,315,361,284[b]
Net Owners equity	78,879,056,590
State capital infusion	10,300,000,000
Total value of company	96,494,417,874

[a] JD 200089.
[b] Value as of October 1, 1949; value as of October 1, 1945 was appraised at 8,211,717,685 old *renminbi*. Final appraised value in 1955 was 6,780,000,000. See JD 200090 and 301201 respectively.

Appendix 5

Shipment of refined salt to selected markets by Jiuda, 1915–1937

(unit: *dan*)

	1916	1917	1918	1919	1920	1921
Tianjin	452	2,103	1,548	1,973	4,476	14,139
Shanghai					1,695	4,371
Qinhuangdao					15	75
Zhangjiakou					157	159
Nanjing		642	18,858	38,987	3,024	1,344
Anqing				4,326	3,696	16,800
Wuhu			1,629	8,284	3,024	10,752
Jiujiang			5,969	26,085	76,748	133,077
Hankou		9,602	35,153	34,530	55,946	86,355
Shashi			5,559	945	28,560	1,344
Yuezhou				5,124	6,006	3,360
Changsha			10,718	9,454	22,740	17,136
Xiangtan				15,400	20,160	15,456
Changde				11,726	16,128	4,032
Total	452	12,348	79,437	156,835	242,377	308,401

	1922	1923	1924	1925	1926
Tianjin	4,740	11,461	16,548	11,522	10,719
Shanghai	5,109	28,924	29,904	18,144	25,200
Qinhuangdao	60	90	60	142	
Zhangjiakou	159	712	154	3,244	
Nanjing	2,691	11,103	5,719	6,384	4,711
Anqing	23,185	29,232	34,944	38,976	15,120
Wuhu	17,811	12,103	9,744	11,760	3,360
Jiujiang	143,490	165,333	137,774	18,837	11,774
Hankou	143,479	129,360	131,726	150,192	127,680
Shashi	20,496	18,480	19,919	10,416	13,776

Continued

	1922	1923	1924	1925	1926
Yuezhou	7,056		11,760	11,088	11,088
Changsha	28,224	10,080	23,184	39,312	36,960
Xiangtan	6,048	3,024	11,760	31,554	20,160
Changde	7,056	1,008	7,056	6,384	8,736
Total	409,604	420,912	440,252	357,955	289,284

	1927	1928	1929	1930	1931
Tianjin	7,077	2,695	4,056	4,032	7,224
Shanghai	10,108	8,400	27,552	33,600	45,360
Qinhuangdao	150		75		
Zhangjiakou					
Nanjing	1,344	4,368	7,728	4,032	4,302
Anqing		16,800	18,480	15,120	26,880
Wuhu		5,040	10,080	3,360	3,528
Jiujiang	52,080	73,934	31,920	43,680	50,400
Hankou	173,040	177,072	107,520	82,320	58,800
Shashi	8,400	11,760	11,760	5,040	8,400
Yuezhou	3,360	3,360	13,440	1,680	8,400
Changsha	16,800	13,440	25,200	16,800	23,520
Xiangtan	3,360	6,720	10,080	13,440	16,800
Changde	3,360	3,360	6,720	1,680	6,720
Bangbu			10,080	2,688	
Wuxi				1,848	3,360
Total	279,080	326,949	284,691	229,320	263,424

	1932	1933	1934	1935	1936	1937*
Tianjin	8,064	23,601	13,104	19,881	28,220	17,220
Shanghai	36,960	53,760	47,880	72,620	57,314	28,890
Nanjing	7,560	11,088	9,912	23,877	20,596	8,560
Anqing	45,360	36,960	6,720	81,313	43,482	21,614
Wuhu	9,072	9,408	6,384	120,605		53,026
Jiujiang	53,760	120,947	67,200	160,707	141,020	32,100
Hankou	107,520	114,240	109,200	119,254	115,182	69,073
Shashi	4,200	5,880		1,712		
Wuyue					8,020	
Yuezhou	6,720	15,120	6,720	10,700	7,040	
Changsha	30,240	50,400	50,400	31,820	14,980	
Xiangtan	15,120	26,880	21,840	34,100	12,540	
Changde	5,040	8,400	18,480	83,453	21,760	
Bangbu				5,136	12,840	17,120
Zhengjiang	2,352	9,072	9,744	19,176	5,778	
Wuxi	3,360	3,360	5,376	25,436	13,268	4,708
Hongchenqiao				798		
Total	335,326	489,144	372,960	810,592	494,960	273,180

*First seven months.

Appendix 6

List of registered salt refineries[a]

Company	Location	Launch Year	Paid-up Capital (yuan)	Production Capacity (dan)
Jiuda	Tanggu	1914	50,000	30,000
			2,100,000	*1,050,000*
Tongyi	Yantai	1919	250,000	60,000
			420,000	*400,000*
Lequn	Yangzhou	1919	300,000	na
Fuhai	Yingkou	1921	100,000	30,000
				200,000
Tongda	Tangfang	1921	179,000[b]	30,000
				50,000
Liyuan	Yingkou	1923	250,000	30,000
				240,000
Huafeng	Yingkou	1923	130,000	30,000
			200,000	*200,000*
Yongyu	Qingdao	1923	800,000[c]	180,000
				400,000
Fengtian	Yingkou	1926	140,000	30,000
			300,000	*200,000*
Yuhua	Yingkou	1927	100,000	30,000
				150,000
Minsheng	Dinghai	1927	50,000	30,000
			70,000	*50,000*
Hongyuan	Fuxian	1928	100,000	30,000
				200,000
Wuhe	Shanghai	1928	100,000	30,000
			240,000	*120,000*
Dinghe	Yuyao	1928	50,000	30,000
			150,000	*50,000*

[a] Caizhengbu yanwuchu, comp.,(1930). Figures in italics from *ZGYZSL* (1933).
[b] Registered capital: 500,000 *yuan*.
[c] Registered capital: 3,000,000 *yuan*.

Appendix 6

List of registered salt refineries

Appendix 7

Market shares of major refineries[a]

I: Hankou: 2.7 *yuan* per *shima dan* and net profit of 0.60 on projected sales of 500,000 *dan*.

	Initial	Mar. 1, 1930	Dec. 5, 1930	% of Total at the End of 1930	% of Total (Apr. 1, 1933)
Jiuda	16.5	19.5		18.30	23.3
Tongyi	16	16.5		15.49	17.3
Fengtian	10.5	9.5		8.9	6.6
Yongyu	10.5	11		10.32	9.1
Liyuan	10	10		9.38	7.6
Huafeng	7.5	7		6.57	5.5
Fuhai	7.5	7		6.57	6
Hongyuan	7.5	7		6.57	6
Wuhe	6.5	7		6.57	4.2
Tongda	6.5	5.5	6	5.63	4.6
Minsheng			6	5.63	4.4
Yuhua					5.1
Total			112	100	100

II: Jiujiang: 2.7 *yuan* per *shima dan* and net profit of 0.65 on projected sales of 300,000 *dan*.

Jiuda	17	17	15.18
Tongyi	17	17	15.18
Fengtian	10.5	10	8.92
Yongyu	10.5	12	10.71
Liyuan	10.5	10	8.92
Huafeng	7.5	7	6.25
Fuhai	7	7	6.25
Hongyuan	7	7	6.25

Continued

Jiuda	17	17	15.18
Wuhe	6.5	7	6.25
Yuhua	5	6	5.35
Tongda		6	5.35
Minsheng		6	5.35
Total		112	100

III: Anqing/Wuhu: 2.6 yuan per *shima dan* and net profit of 1 yuan on projected sales of 100,000 *dan*.

	Initial (Mar. 1, 1930)			Percent of Total	Anqing May 20, 1931	Wuhu May 20, 1931
	Anqing	Wuhu				
Jiuda	37	28	65	41.94	27	26
Yongyu					7	11
Tongyi	19		17	10.96	21	17
Fengtian	13		9	5.8	7	9
Liyuan	15	10	20	12.9	10	10
Huafeng	8		6	3.87	7	6
Fuhai	8		6	3.87	7	6
Wuhe		9	18	11.63	7	9
Hongyuan		8	14	9.03	9	6
Yuhua					8	
Total	100		155	100	110	100

IV: Hunan: 2.7 yuan per *shima dan* and net profit of 0.7 on projected sales of 100,000 *dan*.

	Initial	Mar. 1, 1930	Percent of total	May 19, 1931
Jiuda	24	27	35.53	30
Tongyi	20	21	27.63	16
Liyuan	16	16	21.05	11
Wuhe		12	15.79	8
Hongyuan				7
Fengtian				7
Fuhai				7
Huafeng				7
Yuhua				7
Total		76	100	100

[a] *Source*: Meeting minutes of Joint Office of Salt Refineries in *JY*.

Appendix 8

State-set market share of salt refineries, 1935

(unit = *dan*, or 0.787 *shima dan*)

	Jiuda	Tongyi	Yongyu	Wuhe	Tongda	Minsheng	Total
Hunan	77,000	57,000	35,000	16,000	9,000	8,000	202,000
Hubei	195,000	167,000	76,000	52,000	23,000	23,000	536,000
Jiangxi	150,000	150,000	69,000	44,000	20,000	20,000	453,000
Anhui	86,000	45,000	12,000	11,000	—	—	154,000
Jiangsu[a]	76,000	42,000	10,000	16,000	11,500	12,500	168,000
Others[b]	13,700	10,700	5,000	13,400	—	—	42,800
Total	597,700	471,700	207,000	152,400	63,500	63,500	1,555,800

[a] Shanghai, 138,600; Wuxi, 11,200; Zhenjiang, 18,200; and Nanjing.
[b] Tianjin, 16,500; Qingdao, 3,100; Jinan, 400; Yantai, 100; Bangbu, 14,100; and Hangzhou, 1,000.

Appendix 9

Market share of refined salt in four Yangzi provinces, 1917–1934

Unit = *dan*

Year	LH Q.(Q)[a]	T. Ref.[b](R)	R/Q	Jiuda/Yongyu[c]		Jiuda[d]/(R)
1917		9,615[e]			9,602	(99.8)
1918		68,233			59,028	(86.5)
1919		116,358			115,875	(99.6)
1920		241,456			233,008	(96.5)
1921		303,812[f]			288,312	(94.9)
1922		486,411			396,895	(81.6)
1923		591,143			362,243	(61.3)
1924		675,603			387,417	(57.3)
1925		745,830		382,359(51.3)	369,255	(49.5)
1926		795,587		354,494(44.6)	248,654	(31.3)
1927		853,738		401,560(47.0)	260,400	(30.5)
1928		636,192		338,366(53.2)	311,486	(48.9)
1929	3,344,003	884,724	26.4		273,840	(30.9)
1930		1,219,384		325,584(26.7)	185,808	(15.2)
1931	7,689,000	1,227,000	15.9	352,632(28.8)	233,856	(19.1)
1932	6,999,000	1,116,289	15.9	428,566(38.3)	335,328	(30.0)
1933	6,802,000	1,204,989	17.7	586,588(48.6)	489,111	(40.6)
1934	6,859,000	1,152,719	16.8	588,446(51.0)	458,297	(39.8)

[a] Statutory quota for Anhui (Wuhu and Anqing); Hunan (Changsha, Yueyang, Xiangtan, and Changde); Hubei (Hankou, Yichang, and Shashi); and Jiangxi (Jiujiang) stood at 4,520,000 *dan* in 1929. See Caizhengbu Yanwuchu comp., (1930), sales statistics, 85. Actual sales figures for 1929 from ibid., 93; and for 1931–1934 from *ZGJDYWSZL* IV: Table 2, 257–58.
[b] Total refined salt imported into the division.
[c] Share of the refined salt (R) market supplied by Jiuda and Yongyu.
[d] Jiuda's shipment statistics from Appendix 2.
[e] Figures for 1916–1922 from Adshead (1970), 182.
[f] Figures for 1921, 1923–1926 compiled from Caizengbu, et al., comp. (1934), II, Liaoning, 108–18; Shandong, 120–21; Changlu, 64–5.

Notes

Introduction

1. Fan Xudong, "Jiuda diyige sanshi'nian," [The First Thirty Years of Jiuda] *Haiwang* [Neptune] 17.3 (1944), 18.
2. Patrick Brodie, *Crescent over Cathay* (Hong Kong: Oxford University Press, 1990), 132. On Fan's career, see Zhang Tongyi, *Fan Xudong zhuan [A Biography of Fan Xudong]* (Changsha: Hunan renmin chubanshe, 1987); and Chen Xinwen, *Fan Xudong* (N.p., 2002).
3. Bian Jintao, "Jiuda jingyan gongchang fangwenji," [A Visit to the Jiuda Salt Refinery] *Wenhua jianshe* [Cultural construction] 2.6 (1936), 81–86.
4. Chen Diaopu, "Eulogy of Fan Xudong," in "Commendation file of Fan Xudong" dated November 21, 1945, in the Archives of the National Government (headquarters) 98 held at *AH*; and *Haiwang* 6.19 (1934); 7.1 (1934). The role of nationalism as a motivational force has been an overlooked issue in the study of economic development. See Liah Greenfeld, *The Spirit of Capitalism* (Cambridge, MA: Harvard University Press, 2001), 10.
5. JD 300774, Fan's letter to Fu Bingzhi dated May 31, 1941.
6. The classic treatment of early industrialization and the role of private and bureaucratic/state enterprise remains Liu Kwang-ching, *Anglo-American Steamship Rivalry in China, 182–1874* (Cambridge, MA: Harvard University Press, 1962); and his collected essays on the subject conveniently assembled in Li Zhigang, ed., *Liu Guangjing lun Zhaoshangju [Liu Guangjing's Exposition of China Merchants]* (Beijing: Shehui kexue wenxian chubanshe, 2012).
7. William C. Kirby, "Engineering China: Birth of the Developmental State, 1928–1937" in Yeh Wen-hsin ed., *Becoming Chinese* (Berkeley, CA: University of California Press, 2000), 137–160; Xu Jiansheng, *Minguo shiqi jingji zhengce de yanxi yu bianyi [Continuity and Change of Economic Policy during the Republican Period]* (Fuzhou: Fujian renmin chubanshe, 2006), 35–45.
8. Wang Zhenghua, ed., *Jiang Zhongzheng dang' an shilue gaoben* [Draft of President Jiang Jieshi's abridged archival materials] (Taibei: Academia Historica, 2011-), entry dated July 9, 1929; and Margherita Zanasi, *Saving the Nation: Economic Modernity in Republican China*] (Chicago: University of Chicago Press, 2006), passim.

9. Pierre Laszlo, *Salt,* trans. Mary Beth Mader (New York: Columbia University Press, 2001), 74.
10. For a succinct discussion and the conflicts among these criteria for tax structures, see John F. Due and Ann F. Friedlaender, *Government Finance* (Homewood, IL: Richard D. Irwin, 1973), 233–240; John S. Dzienkowski and Robert J. Peroni, *Natural Resource Taxation* (Durham, NC: Carolina Academic Press, 1988), 19–23.
11. S. A. M. Adshead, *The Modernization of the Chinese Salt Administration* (Cambridge, MA.: Harvard University Press, 1970), passim; Julia Strauss, *Strong Institutions in Weak Polities: State Building in Republican China, 1927–1940* (Oxford: Clarendon Press, 1998), chapters 3 and 4.
12. Zhang Lijie, *Nanjing guominzhengfu de yanzheng gaige yanjiu* [A Study of Salt Administration Reforms under the Nanjing Nationalist Government] (Beijing: Zhongguo shehui kexueyuan chubanshe, 2011), 17–22.
13. Lloyd Eastman, *The Abortive Revolution* (Cambridge, MA: Harvard University Press, 1974), passim; Parks Coble, *The Shanghai Capitalists and the Nationalist Government, 1927–1937* (Cambridge, MA: Council on East Asian Studies, Harvard University, 1980), passim; Richard C. Bush, *The Politics of Cotton Textiles in Kuomintang China* (New York: Garland Press, 1982), 16. For a summary of various positions, see Margaret M. Pearson, *China's New Business Elite* (Berkeley, CA: University of California Press, 1997), 55, fn. 33.
14. Thomas Rawski, *Economic Growth in Prewar China* (Berkeley, CA: University of California Press, 1989), 116.
15. Wellington K. K. Chan, "The Organizational Structure of the Traditional Chinese Firm and its Modern Reform," *Business History Review* LVI: 2 (1982), 218–235; Wang Siu-lun, "The Chinese Family Firm: A Model," *The British Journal of Sociology* 36.1 (1985), 58–72; Tim Wright, "The Spiritual Heritage of Chinese Capitalism: Recent Trends in the Historiography of Chinese Enterprise Management," in Jonathan Unger, ed., *Using the Past to Serve the Present: Historiography and Politics in Contemporary China* (Armonk, New York: M.E. Sharpe, 1993), 205–238.
16. Chen Zhongping, *Modern China's Network Revolution: Chambers of Commerce and Sociopolitical Change in the Early Twentieth Century* (Stanford, CA: Stanford University Press, 2011), passim; S. Gordon Redding, *The Spirit of Chinese Capitalism* (Berlin: Walter de Gruyter, 1990), 4.
17. Kenneth Pomeranz, "'Traditional' Chinese Business Forms Revisited: Family, Firm and Financing in the History of the Yutang Company of Ji'ning," *Late Imperial China* 18.1 (1997), 1–38; Zou Jinwen, "Lun Zhongguo jindai minying gufen qiye de jiazu teshi," [On the Family Character of Private Corporations in Modern China] *ZGJJSYJ* 1 (2004), 18–27; Dai Yifeng, "Chuantong yu jindai," [Tradition and Modern] *ZGSHJJSYJ* 1 (2004), 65–73; Sherman Cochran and Andrew Hsieh, *The Lius of Shanghai* (Cambridge, MA: Harvard University Press, 2013), passim.
18. For American and European examples, see Yoram Ben-Porth, "The F-connection: Families, Friends, and the Organization of Exchange," *Population Development Review* 6 (1980), 1–30; Harold James, *Family*

Capitalism (Cambridge, MA: Harvard University Press, 2006), passim; Paul Windoff, *Corporate Networks in Europe and the U.S.* (Oxford: Oxford University Press, 2002), 12–13; Naomi R. Lamoreaux, Daniel M. G. Raff, and Peter Temin, "Beyond Markets and Hierarchies: Toward a New Synthesis of American Business History," *NBER Working Paper* No. 9029 (July, 2002), passim; Kenneth Lipartito, and David B. Sicilia, eds., *Constructing Corporate America* (New York: Oxford University Press, 2005), passim; Pamela Walker Laird, *Pull* (Cambridge, MA: Harvard University Press, 2006), passim. For a critique of the scholarship on Chinese networks as a form of orientalism, see Arif Dirlik, "Critical Reflections on "Chinese Capitalism as Paradigm," *Identities* 3.3 (1997), 303–330. The literature on Asian business network is far too big to be cited in full here. See Mark Fruin, ed., *Networks, Markets, and the Pacific Rim* (New York: Oxford University Press, 1998), passim; and Nicole W. Biggart and Gary G. Hamilton, "On the Limits of a Firm-based Theory to Explain Networks: Western Bias of Neoclassical Economics," in Nitin Nohria and Robert G. Eccles, eds., *Networks and Organizations* (Cambridge, MA: Harvard Business School, 1992), 471–490.

19. I owe this formulation to the comparative business system approach pioneered by Richard Whitley, *Divergent Capitalisms* (Oxford: Oxford University Press, 1999), chapter 1.
20. On the tradition to modern model, see, for example, Wellington Chan, "Tradition and Change in the Chinese Business Enterprise," in Robert Gardella, Jane K. Leonard, and Andrea McElderry, eds., *Chinese Business History* (New York: M.E. Sharpe, 1998), 127–144; Liu Wenbin, *Jindai Zhongguo qiyue guanli sixiang yu zhidu de yanbian* [The Thought and Institutional Transformation of Management in Modern China] (Taibei: Academia Historica, 2001), passim; and Wang Chuhui, *Zhongguo jindai qiyue zuzhi xingtai de bianqian* [Transformation of Business Organization in Modern China] (Tianjin: Renmin chubanshe, 2001), passim., esp. 301. On the impact of economic ideology and national culture on business organization, see David A. Ralston, "The Crossvergence Perspective: Reflections and Projections," *Journal of International Business Studies* 39.1 (2008), 27–40.
21. Hilton L. Root, *Capital and Collusion* (Princeton: Princeton University Press, 2006), 246; Peter Z. Grossman, ed., *How Cartels Endure and How They Fail* (Cheltenham, UK: Edward Elgar, 2004), 1.
22. Charles Perrow, "Market, Hierarchies, and Hegemony: A Critique of Chandler and Williamson," in Andrew H. Van de Ven and William F. Joyce, eds., *Perspectives on Organizational Design and Behavior* (New York: Wiley, 1981), 371–86; Hans B. Thorelli, "Networks: Between Markets and Hierarchies," *Strategic Management Journal* 7.1 (1986), 37–51; Walter W. Powell, "Neither Market nor Hierarchy: Network Forms of Organization," *Research in Organizational Behavior* 11 (1990), 295–336; Grahame F. Thomson, *Between Markets and Hierarchy: the Logic and Limits of Network Forms of Organization* (Oxford: Oxford University Press, 2003), 21–52.
23. Adam Smith, *The Wealth of Nations* (New York: Random House Modern Library ed., 1937), 423.

24. On this debate, see Rogers Brubaker, "In the Name of the Nation: Reflections on Nationalism and Patriotism," in Philip Abbott, ed., *The Many Faces of Patriotism* (Lanham, MD: Rowman and Littlefield, 2007), 37–52; Mary G. Dietz, "Patriotism: A Brief History of the Term," in Igor Primoratz, ed., *Patriotism* (Amherst, NY: Humanity Books, 2002), 201–217.

1 The Institutions

1. Ding Changqing, ed., *Minguo yanwu shigao* [A Draft History on Republican Salt Affairs] (Beijing: Renmin chubanshe, 1990), 113; Adshead (1992), 337; and Julia Strauss, "Rethinking Institutional Capacity and Tax Regimes," in Deborah Bräutigam, Odd-Helge Fjeldstad, and Mick Moore, eds., *Taxation and State-Building in Developing Countries* (Cambridge: Cambridge University Press, 2008), 212–4.
2. Esson M. Gale, *Salt for the Dragon* (East Lansing, MI: Michigan State College Press, 1953), 98, 129, 131, 134. See also Lin Meili, *Xiyang shuizhi zai jindai Zhongguo de fazhan* [The Development of Western Taxation Systems in Modern China] (Taibei: Academia Sinica, 2005), 32.
3. Adam Smith, *Yuan Fu* [*The Wealth of Nations*] trans. Yan Fu (Shanghai: Nanyang gongxue yishuyuan, 1902 ed.; repr. *Xuxiu Siku quanshu ed.*, Guji chubanshe, 1995–1999), v. 1297, 723. See also Xia Guoxiang, "Jindai Zhongguo yanzheng gaige sixiang chutan," [Perspectives on Gabelle Administration Reform in Modern China] *YYSYJ* 3 (2009), 40–6.
4. Gong Jinghan, "Yan gui diding buzhi shoushuiyi," in He Changning, ed., *Huangchao jingshi wenbian* [Anthology of Essays on Statecraft] (1903 repr. ed.), *juan* 49, 29b–30a; Zhang Yuanding, *Zhenxiu Hedong yanfa beilan* [Updated Compendium on Salt Administration of Hedong Division] (preface dated 1882), *juan* 5, 2a–27a; Tang Shouqian, "Weiyan" [Cautionary Words] in Zhengxie Zhejiangsheng Xiaoshanshi weiyuanhui wenshi gongzuo weiyuanhui, ed., *Tang Shouqian shiliao zhuanji* [Historical Materials on Tang Shouqian] (N.p., n.d., postscript dated 1992), 240–42; and Yanmi (Jing Benbai), "Sishi nianlai yanwu geming zhi zhong jiantao," [Forty Years of Revolutionary Changes in Salt Affairs: A Summary Review] in *YMZK* (1935), 3.
5. Liu Jinchao, comp., *Qingchao xu wenxian tongkao* [Compendium of Historical Records of the Qing Dynasty, continued] 4 vols. (Shanghai: Shangwu yinshuguan ed., 1936), v. 1, *juan* 34, *kao* 7871.
6. Cang Jiang (Liang Qichao), "Gai yanfayi," [A Proposal for Salt Administration Reform] in *Guofengbao* [National review] 1.5 (1910), 1–5; 1.6 (1910), 1–15.
7. Xiong Xiling's telegram to Tan Yankai ca. June 3, 1912, in Lin Zengping and Zhou Qiuguang, eds., *Xiong Xiling ji* [Collected Works of Xiong Xiling] 2 vols. (Changsha: Hunan renmin chubanshe, 1985), I: 342–3.
8. Zhang Jian, "Gaige quanguo yanzheng jihuashu," [A Plan for National Gabelle Reform] in *YZCK* (1921), I: 123–59. See also Watanabe Atushi, "Shimmatsu ni okeru Zhang Jian no enhōron to sono rekishi teki haikei,"[Chang Jian's

Plan in Salt Reform and its Historical Background in Late Qing] *Kumamoto daigaku kyōiku gakubu kiyō* 24 (1975), 13–44; Adshead (1992), 322; Qiu Tao, "Zhang Jian de yanye gaige shixiang he shijian," [Theory and Practice of Zhang Jian's Salt Reform] *Qingshi luncong* [Studies on Qing History] (2000), 172–192; and Liu Changshan, *Qingdai houqi zhi minguo chunian yanwu de bianqe* [Reforms of the Salt Administration from Late Qing to the Early Republic] (Taibei: Wenshizhe chubanshe, 2007), 140–1.
9. Zhang Jian's memorandum to Huang Xing dated February 10, 1912, in Zhongguo di'er lishi dang'anguan comp., *Zhonghua minguo dang'an ziliao huibian* [Collected Archival Materials on Republican China] Series 2 (N.p: Jiangsu guzhi chubanshe, 1991), 366–7. See also Watanabe Atsushi, "Shingai kakumeiki ni okeru ensei kaikaku undō," [On the Salt Reform Movement during the 1911 Revolution] *Kumamoto daigaku kyōiku gakubu kiyō* 27 (1978), especially 71–5.
10. Yang Shoutong, *Yunzai shanfang luigao* [Brief Notes by Category from the Mountain House amidst Clouds] (N.p: n.d.), *juan* 2, 5a–b; Jia Shiyi, *Minguo caizheng shi* [A Fiscal History of the Republic] 2 vols. (Shanghai: Shangwu yinshuguan, 1917), I: 25–38; Qian Jiaqu, "Zuijin sanshinian de Zhongguo caizheng," [National Finance in the Last Thirty Years," *Dongfang zazhi* [*Eastern Miscellany*] 31.1 (1934), 115; and Watanabe Atushi, "Shimmatsu ni okeru ensei no chūō shūkenka seisaku ni tsuite" [On the Centralization Policy of Salt Administration in Late Qing], in Nakajima Satoshi senshi koki kinen jigyōkai ed., *Nakajima Satoshi sensei koki kinen ronshu* [Studies in Asian History Dedicated to Professor Satoshi Nakajima on His Seventieth Birthday], 2 vols. (Tokyo: Kyuko shoin, 1980), II: 657–80.
11. Lin Zengping and Zhou Qiuguang, eds. (1985), I: 319, 326, 528.
12. Sun Yat-sen telegram to Li Yuanhong, second month of 1912, in Zhongyang dangshi shiliao bianxie weiyuanhui comp., *Guofu quanji* [Complete Works of Sun Yat-sen] (Taibei: Zhongyang wenwu gongyingshe, 1947), 202.
13. On the historiography of the Reorganization Loan, see S. A. M. Adshead (1970), passim., and K. C. Chan, "British Policy in the Reorganization Loan to China 1912–1913," *Modern Asian Studies* 5.4 (1971): 355–72.
14. On November 13, 1911. See Frank H. H. King, *The Hongkong Bank in the Period of Imperialism and War, 1895–1918* (Cambridge: Cambridge University Press, 1988), 477; and Roberta Allbert Dayer, *Finance and Empire* (New York: St. Martin's Press, 1988), 65. Guangdongsheng jiexue shehui kexue yanjiusuo lishi yanjiushi, et al., comp., *Sun Zhongshan nianpu* [A Chronological History of Sun Yat-sen] (Beijing: Zhonghua shuju, 1980), 126 listed the date of meeting under November 10, 1912.
15. See, for example, a six million mark loan from Carlowitz & Co. for Krupp artillery by the Nationalist-controlled government in Zhejiang. Xu Yisheng, *Zhongguo jindai waizhaishi tongji ziliao* [Statistics on Modern China's Foreign Debt, 1853–1927] (Beijing: Zhonghua shuju, 1962), 115–7 and Shen Xiaomin, *Chuchang yu qiubian* [Between Continuity and Change] (Beijing: Sanlian shudian, 2005), 89.
16. Alston to Balfour dated September 12, 1917, in *FO* 371.2908. See also David McLean, "British Banking and Government in China: the Foreign

Office and the Hongkong and Shanghai Bank, 1895–1914" (PhD dissertation, Cambridge University, 1973), 129; and Foreign Office to Hongkong Bank dated March 12, 1912, in *ZGJDYWSZL* I: 13.
17. Zhou Xuexi's offer to the groups consortium dated August 20, 1912; and Hillier's rejection letter September 22, 1912, in *ZGJDYWSZL* I: 49–53.
18. Gale (1953), 133, 144.
19. John V. A. MacMurray comp. and ed., *Treaties and Agreements with and Concerning China*, 2 vols. (New York: Oxford University Press, 1921), 2: 1013, 1020; and Xu Yisheng (1962), 109–10.
20. MacMurray (1921), 2: 1007–15.
21. See Roberta A. Dayer, *Bankers and Diplomats in China* (London: Frank Cass, 1981), 27; and Arthur S. Link et. al., eds., *Papers of Woodrow Wilson* (Princeton: Princeton University Press, 1978), v. 27, 192–3.
22. William N. Goetzmann and Andrey Ukhov, "China and the World Financial Markets, 1870–1930: Modern Lessons from Historical Globalization," *The Wharton School Financial Institutions Center Working Paper* (2001), 3; and King (1988), 2: 503.
23. Jordan to W. Langley dated June 13, 1916, in FO 350.15.
24. *The Times*, May 1, 1914; Richard Dane to Bank Group dated July 18, 1917, in HSBC SHG 319.3, p. 1898; and A. N. Young, "Confidential Memorandum on the Position of the Salt Revenue Administration," dated February 20, 1942 in Box 74, A. N. Young Papers held at the Hoover Institution.
25. MacMurray (1921), 2: 1010.
26. On Zhang, see Elisabeth Köll, *From Cotton Mill to Business Empire* (Cambridge, MA: Harvard University Asia Center, 2003), passim. On Zhou, see Zhou Xiaojun, *Zhou Xuexi chuanji huibian* [Biographies of Zhou Xuexi] (Lanzhou: Gansu wenhua chubanshe, 1997), passim.
27. *ZGJDYWSZL* I: 115–6.
28. Letter of transmittal to the parliament dated November 10, 1912, in Dier lishi dang'anguan comp., *Zhonghua minguoshi dang'an ziliao huibian* [Archival Materials on the History of Republican China] (Nanjing: Jiangsu guzhi chubanshe, 1991) series 3, finance, 2: 1361–6; and Zuo Shuzhen, *Zhongguo xianjin yanwu gaiyao* [A Concise Summary of Salt Administration in China Today] (Beijing: Yanwu xuexiao, n.d.), 49a.
29. Ben Bai (Jing Benbai), "Yanwu tihesuo quanfei wenti," [Keeping or Abolishing the Revenue Inspectorate?], repr. in *YZCK* (1931), II, 1: 242.
30. Liang Qichao, "*Yanzheng zazhi* xu," [Preface to the *Journal on Salt Administration*], repr. in *YZCK* (1921) I, 1: 3–4.
31. Adshead (1970), 210–215; and Julia Strauss (1998), chaps. 3 and 4.
32. British government memorandum dated June 27, 1912; Addis telegram to Hillier dated September 11, 1912 and Ishūin's confidential telegram to Tokyo dated July 13, 1912 in *ZGJDYWSZL* I: 30–31, 34–37, 45; and McLean (1973), 199.
33. MacMurray (1921), 2: 1026–9.
34. For an analysis of the regulations and the basis of Dane's authority, see *ZGJDYWSZL* I: 141–53; and Chen Zhenping, "Minguo chunian de yanwu

gaige," [Salt Administration Reforms in the Early Republican Period] *ZGJJSYJ* (1994), supplemental issue, 125–34.
35. Adshead (1970), 56, 92–4; and Mark Kurlansky, *Salt: A World History* (New York: Penguin Books, 2002), 334–43.
36. Ding An [Richard Dane], *Gaige yanwu baogaoshu* [Report on the Reform of the Salt administration] (Beijing: Yanwushu, 1922), 1: 68b.
37. Adshead (1970), 92; and Liu Foding, "Lun Zhongguo yanwu guanli de xiandaihua," [On the Modernization of China's Salt Administration], in Peng Zeyi and Wang Renyuan, eds., *Zhongguo yanyeshi guoji xueshu taolunhui lunwenji* [Proceedings of the International Symposium on the History of Salt in China] (Chengdu: Sichuan renmin chubanshe, 1991), 143–58.
38. Zhang Dazhuang, "Beiyang junfa shiqi Beijing yanwushu chengli chuqi de qingkuang," [The Early Years of the Salt Administration in Beijing under the Northern Warlords] in Zhongguo renmin zhengzhi xieshang huiyi, eds., *Shanghai wenshi ziliao quangao* [Extant Manuscripts on Historical Materials in Shanghai] (Shanghai: Guji chubanshe, 2001), IV: 135–41.
39. Gale (1953), 126. Gale joined the service in 1914.
40. Letter to L.C. Wolcott dated August 6, 1927, in R.D. Wolcott Letters held at Bentley Historical Library, University of Michigan, Ann Arbor. While in the service, Wolcott also reported to the U.S. Army Intelligence since 1923, if not earlier.
41. John Jordan's letters to W. Langley dated April 16, 1916, and May 23, 1916; Max Fuller to Jordan dated April 4, 1919, in Jordan Papers, *FO* 350.15 and 16; and Sir Louis Dane (Sir Richard's brother) to Lord Hardings dated August 18, 1917 in "Political Correspondences, China," *FO* 371.2915.
42. "Yanshuifa cao'an" [Draft gabelle law], reprinted in *YZCK* (1921), I: 273–282; letter of transmittal to the parliament dated October 20, 1913, in Zhongguo di'er dang'anguan, comp. (1991), series 3, caizheng 2, 1372–3. This might have standardized the gabelle on a per *dan* basis for the state, but did not solve the problem of varying local weight standards and hence the effective gabelle rate at the retail level. See chapter 6.
43. Guojia shuiwu zongju, ed., *Zhonghua minguo gongshang shuishou shi—yanshui juan* [A History of Republican Tax—Gabelle] (Beijing: Zhongguo caizheng jingji chubanshe, 1999), 21–2, 46–7.
44. Benbai, *YZCK* (1931), 1: 241–5.
45. Dane's memo to Zhang Hu dated November 14, 1914, in *ZGJDYWSZLBY*; Ding An (1922), 2: 65b, 67a–70a; *DGB* November 12,1914; and Zhang Tongli, "Zhang Hu de yisheng," [The Life of Zhang Hu] *TJWSZL* 23 (1983), 152. Tongli was Hu's son.
46. Ding An (1922), 3: 157b; Zhang Tongli (1983), 152.
47. Xu Zhucheng, "Li Sihao shengqian tan congzheng shimo," [Li Sihao's Reminiscence on Politics], *Wenshi ziliao xuanji* [Selected Materials on History and Literature, Shanghai] 22 (1978), 158–9. For the charters and contracts of the company in 1915, see *CLHB* 680.7.1279 and 1413; Guojia shuiwu zhongju, ed. (1999), 45; and Hua Kege, "Changlu yanwu fengchao

zhong de shida leishang an," [The Case of Ten Bankrupted Changlu Merchants] in Ding Changqing ed., *Jindai Changlu yanwu* [Salt Affairs of Modern Changlu] (Beijing: Zhongguo wenshi chubanshe, 2001), 130.
48. Li Yeting, "Yuji yanwu gongsi xingshuai gailue," [The Rise and Fall of the Yuji Salt Company] *TJWSZL* 26 (1984), 122–6; and Qi Xiemin, "Cong Dexing yanwu gongsi de xingshuai kan Huabei zhengzhi de xiaochang," [Shifting Politics in North China as Seen from the Rise and Fall of the Dexing Salt Company] *TJWSZL* 95 (2002), 192–8.
49. See, for example, petitions of Jinyinhang, Tongxingwen and Jiyuan dated from 1912 to 1920 for relief in *CLHB* 680.7.68.
50. Gale (1953), 129; Ding An (1922), 2: 79b, 85a–b.
51. Ding An (1922), 2: 89a–b, 93a–94a; 3: 175b.
52. Dane's letter to Hillier dated April 30, 1914, in *ZGJDYWSZL* I: 201.
53. F. A. Cleveland memo to Song Ziwen dated November 1–3, 1932, in Box 32, Arthur N. Young Papers, held at the Hoover Institution.
54. Yang Shoutong, comp., *Changlu yanzheng jiyao* [Essentials of Changlu Salt Administration] (n.p.: preface dated 1913), 16a–22b; Cheng Xuezhen, *Changlu yanwu gaige shikuanglu* [A Veritable Record of Reforms in the Changlu Division] (n.p., preface dated 1917), 15b–17a; and *ZGYZSL* I.1: 9.
55. Gabelle in arrears due to Beijing before the signing of the Reorganization Loan apparently were not audited, nor under the control of the inspectorate after 1914. This made Li Sihao's life much easier as finance minister. See Xu Zhucheng (1978), 149.
56. *YZCK* (1921), I: 276–7.
57. This is contrary to the claim that rate increases did not contribute to the growth of Changlu gabelle revenue for those years. See Adshead (1992), 341 and Chen Zhenping (1994), 136.
58. Cheng Xuezhen (preface dated 1917), 17; and *ZGJDYWSZL* I: 308–26.
59. Reginald Gamble to Pan Fu dated September 1920 in *ZGJDYWSZLBY*. State operation, an alternative dismissed by Dane, also generated a sizable portion of the growing salt revenue for the central government, ranging from 8,193,417 yuan in 1912 to 2,435,979 for 1914, 4,479,493 *yuan* for 1916, and 5,465,661 for 1919. See Zhongguo dier lishi dang'anguan, comp. (1991), Series 3, caizheng, 1: 291, 310, 336, 598.
60. Ding An (1922), II: 94b–95a; and *ZGJDYWSZL* I: 321–4.
61. Draft of revised gabelle law dated October 8, 1915 and promulgated March 9, 1918 in Zhongguo di'er dang'anguan, comp. (1991), series 3, caizheng 2: 1390–1392; 1394–1396; and Dane's comments in Ding An (1922), 4: 235.
62. Guojia shuiwu zongju, ed. (1999), 138.
63. F. A. Cleveland memo to Song Ziwen dated June 17, 1931 in Box 32, Arthur N. Young Papers.
64. Wang Zhong, "Yuan Shikai tongzhi shiqi de yanwu he yanwu gaige," [Salt Administration and Reforms under Yuan Shikai's Rule] *Jindaishi yanjiu* [Studies in Modern History] 4 (1987), 95–121; Liu Jinghua, "Minguo chuqi ge dayanqu gaige zhixiao fenxi," [An Analysis of the Accomplishments of Reforms at the Major Salt Divisions during the Early Republic] *ZGJJSY* 4 (2002): 86–94; and id., "Minguo chuqi shiyan maoyi ziyouhua lunxi," [An

Analysis of Liberalization of Alimental Salt Trade during the Early Republic] *ZGSHJJSYJ* 2 (2003): 74–82. On the case of Sichuan, see Madeleine Zelin, *The Merchants of Zigong* (New York: Columbia University Press, 2005), 229–35.
65. Gale (1953), 137.
66. Dane's retirement might also be due to constant interference by the British government and his frustrations at reform, as did his son in 1928. See respectively, John Jordan to W. Langley, *FO* 350.15; Liu Changshan (2007), 199; and letter to L. C. Wolcott dated December 23, 1928, in W. D. Wolcott letters.
67. Beijing to London Head office dated April 29, 1926, in SHG II 319.7, *HSBC*; Gale (1953), 130; and Adshead (1970), 197–8. For an accounting of the withholdings, rising administrative cost, and the declining share of gabelle to the central government, see *ZGJDYWSZL* I: 450; Tables 2–3, *ZGJDYWSZL* IV, 255–270.
68. Liu Cunliang, "Zhongguo renmin zhi yanshui fudan," [The Salt Tax Burden on the Chinese People] *Zhongguo jingji* [The Chinese Economy] 2.2 (1934), 1–40; and Guojia shuiwu zongju, ed. (1999), 45, 52–54. On the disintegration of the salt service, see also "Chinese Government Salt Revenue Administration Correspondences," SHG II 319.2–7, *HSBC*.

2 The Networks

1. Fan Xudong's *curriculum vitae* in "Junshi weiyuanhui weiyuanchang shicongshi renshidang," [Generalissimo's Office Personnel Files] 12957 held at *AH*.
2. Not at MIT as identified in S. A. M. Adshead (1991), 171.
3. Xiong served as the superintendent of the provincial academy, and, as an aide to Zhao Erxun, worked with Yuanlian as early as 1905. From February to December 1914, Liang was director of the bureau.
4. Ben Bai (Jing Benbai), "Jiuda jingyan gongsi chuanglishi," [A History of the Jiuda Salt Refinery] *YMZK* 1 (1935), *shiliao* [historical materials] II, 4.
5. On the limitations and achievements of Republican China's company law, development of corporate forms, and guaranteed dividend in modern China, see respectively, William C. Kirby, "China Unincorporated: Company Law and Business Enterprise in Twentieth Century China," *Journal of Asian Studies* 54.1 (1995), 43–63; Zhang Zhongmin, "Jindai Zhongguo gongsi zhidu yanjiu de huigu yu zhanwan," [A Review and Prospects of Studies on the Corporate System in Modern China] in Chugoku kigyoshi kenkyūkai, ed., *Chugoku kigyoshi kenkyū no seika to katai* [Achievements and Issues in Chinese Business History] (Tokyo: Kyuko shoin, 2007), 47–61; and Li Yu, *Beiyang zhengfu shiqi qiye zhidu jiegou shilun* [On the Structure of the Corporate System during the Northern Warlord Period] (Beijing: Shehui kexue wenxian chubanshe, 2007).
6. Greta Krippner, Mark Granovetter, Fred Block, et al., "Polanyi Symposium: A Conversation on Embeddedness," *Socio-Economic Review* 2 (2004),

109–35, esp. 115. I have added common foe to the usual list of nine mutualities. On the five affinities, see Lin Qixian and Lu Liangba, eds., *Wuyuan wenhua gailun* [The Culture of Five Affinities] (Fuzhou: Fujian renmin chubanshe, 2003), passim.
7. In contrast to a simple layered order, a heterarchy is a system of organization replete with overlap, multiplicity, mixed ascendancy, and/or divergent-but-coexistent patterns of relation. The literature of network analysis is too big to cite here. For two useful recent overviews, see Peter R. Monge and Noshir S. Contractor, eds., *Theories of Communication Networks* (Oxford: Oxford University, 2003); and Emanuela Todeva, *Business Networks* (London: Routledge, 2006), 130–60.
8. See, for example, Dwight Perkins, "Law, Family Ties, and the East Asian Way of Business," in Lawrence E. Harrison and Samuel P. Huntington, eds., *Culture Matters: How Values Shape Human Progress* (New York: Basic Books, 2000), 232–43.
9. Aino Halinen and Jan-Åke Törnroos, "Using Case Methods in the Study of Contemporary Business Networks," *Journal of Business Research* 58.9 (2005), 1285–1297; and Julie M. Hite, "Patterns of Multidimensionality among Embedded Network Ties: A Typology of Relational Embeddedness in Emerging Entrepreneurial Firms," *Strategic Organization* 1.1 (2003), 9–49.
10. Martin Gargiulo and Mario Benassi, "The Dark Sides of Social Capital," in Th. A. Roger, J. Leenders, and Shaul M. Gabby, eds., *Corporate Social Capital and Liability* (Norwell, MA: Kluwer Academic, 1999), 298–322.
11. For the lists, see *JDJYZJ* I: 22–26. The December 1915 list totaled 443 shares instead of 446. By the time of company registration, there were 39 shareholders. See Sun Duoti memo to Tao Jiayao dated June 15, 1915 in the Archives of the Changlu Division 680.7.1077 held at *CLHB*. The 1941 shareholder register is held at the Archives of the Ministry of Economic Affairs, 18–23-72-37-002, *AS*.
12. "Jiuda jingyan gongsi lici huiyilu," [Jiuda Salt Refinery Company Meeting Minutes] *YZZZ* 19 (1915), *zalu* [miscellany], 1; and Jing Benbai (1935), 5–7.
13. A protégé of Xiong Xiling, Zhou was forced to resign by Xuexi, no relative, in 1915. See Ji Xiaoquan and Yang Guzhi, "Jincheng yinhang de cangli yu fazhan," [The Establishment and Growth of the Jincheng Bank] in Xu Jiajun et al., eds., *Zhou Zuomin yu Jincheng Yinhang* [Zhou Zuomin and the Jincheng Bank] (Beijing: Zhongguo wenshi chubanshe, 1993), 13.
14. Wang Daxie, a native of Hangzhou, Zhejiang, was a cousin of Kangnian, and also a leader of the provincial railroad protection movement. Jing's elder sister was married to Daxie's nephew Zhongyu. See Yi Huili, "Lun Zhejiang shishen yu Zhelu feiyue" [On the Railroad Protection and Treaty Abolition Movement and Zhejiang Gentry] http://www.chinese-thought.org/shgc/ 002590_2.htm. Daxie was thus a relative by marriage to Jing, but he probably also knew Liang and the Fan brothers as he was the official supervisor of Chinese students studying in Japan in 1902.
15. On Chen, who also worked closely with Liang Qichao in overthrowing Yuan Shikai, see Zhang Yufa, *Min'guo chu'nian de zhengdang* [Political

Parties in Early Republican China] (2nd ed., Taibei: Institute of Modern History, Academia Sinica, 2002), 138–43.
16. The list of Liang's students and friends is compiled from Ding Wenjiang comp., *Liang Rengong xiansheng nianpu chagnbian chugao* [Draft of Liang Qichao's Chronological Biography] (Taibei: Shijie, 1959); and Guoli Zhongyang tushuguan tiecangzhu, comp., *Liang Qichao zhijiao shouza* [Correspondences of Liang Qichao's Alter Ego] (Taibei: Zhongyang tushuguan, 1995); Philip C. C. Huang, *Liang Ch'i-chao and Modern Chinese Liberalism* (Seattle: University of Washington, 1972), 117–33; and Dong Fangkui, *Liang Qichao yu lixian zhengzhi* (Wuhan: Huazhong shifan daxue chubanshe, 1991), 89. Fan Bingjun has also been mentioned as an active member of the Political Views Society (*Zhengwen she*) in Japan, working together with Xu Fosu under Liang.
17. Liang Qichao, *Liang Qichao quanji* [Complete Works of Liang Qichao] (Beijing: Beijing chubanshe, 1999), 5: 2792; 8: 2522.
18. Yuanlian served as minister of education, 1912–13; and 1916–17 concurrently as minister of interior affairs. Although Fan called Yuanlian his elder brother, his commemorative essay for unknown reasons referred to him as a first cousin. See Xing Nantian, "Fan Rui" in Zhongguo shehui kexueyuan Jindaishi ziliao bianjibu, comp., *Min'guo renwu beichuanji* [Collected Commemorative Essays and Biographies from the Republican Period] (Chengdu: Sichuan renmin chubanshe, 1997), 515. On the relationship between Liang and Xudong, see *Liang Qichao jiashu* [Liang Qichao's Family Letters] (Beijing: Zhongguo minlian chubanshe, 2000), 535; and Ren Zhiyuan, "Fan Xudong xiansheng shishi zhounian ji'nian," [Commemorating Mr. Fan Xudong's Passing] *Haiwang* 19.2 (1946), 10.
19. Fan Xudong, "Jiuda diyige sanshinian," [The First Thirty Years of Jiuda] *Haiwang* 17.2 (1944), 10.
20. On Zhang, a fellow Hunanese and graduate of Waseda University, see Zhang Zhizhong, "Shiyejia Zhang Kegong shilue," [A Brief Biography of the Industrialist Zhang Qinji] *Hunan wenshi ziliao* [Hunan Historical Materials] 2 (1989), 78–90.
21. On strong and weak ties in social network analysis, see Mark Granovetter, "The Strength of Weak Ties," *American Journal of Sociology* 78 (1973), 1360–80.
22. Jinlin is a member of the noted industrialist Fang family of Ningbo. Both Jing and Fang graduated in the same class of the civil service examination.
23. For a partial listing of the society's members, see issues of the *YZZZ*. Not all members, however, were shareholders of the company and vice versa.
24. On the careers of these officials and politicians, see Liu Shoulin, Wan Renyuan, Wang Yumin, and Kong Qingtai, et al., eds., *Minguo zhiguan nianbiao* [Chronological Dictionary of Republican Officials] (Beijing: Zhonghua shuju, 1995). Space would not permit me to analyze the shifting political alliances and the resulting "weak" ties among these politicians/shareholders here. See, for example, Xiang Ruikun letters on his cultivation of Liang Qichao to Yuan Shiliang dated April 26, 1911, in Wang Ermin ed., *Yuanshi jiazang jindai mingren shoushu* [Letters of Prominent Figures in

Modern China held by the Yuan family] (Taibei: Academia Sincia, 2001), 569, 574.
25. Jiuda shareholder register, 1941. Zhou was also a major shareholder of various other companies in Fan's group, including Yongli Chemical Industries, Yongyu Salt Refinery, Bohai Chemicals, and Tianjin Steamship Transportation Co., Ltd. See shareholder registers of the various companies held at *AS*.
26. On the relationship between Fan, Zhou, and Jing, see Xu Jiajun et al., ed. (1993), 41–3, 89. Before Zhou left for Japan, he studied as a student and protégé of Luo Zhenyu, brother of Zengfang, another Jiuda shareholder.
27. See Yu Xiaoqiu's manuscript memoir, Q264.1.1216 held at *SM*.
28. *Haiwang* 20.26 (1948), 411; and Yu Xiaoqiu's memoir held at *SM*.
29. For an example of such contracts, see one between Jincheng and Jiuda for one million *yuan* dated November 3, 1936. On the national goods movement, see Quanguo zhengxie wenshiban, et al., eds., *Zhongguo jindai guohuo yundong* [Modern China's National Goods Movement] (Beijing: Zhongguo wenshi chubanshe, 1996); and Karl Gerth, *China Made: Consumer Culture and the Creation of the Nation* (Cambridge, MA: Harvard University Asia Center, 2003), passim.
30. On Wang, see Liao Mei, *Wang Kangnian* (Shanghai: Shanghai Guji chubanshe, 2001), passim.
31. "Ye Kuichu," *Gongshang jingji shiliao congkan* [Anthology of Historical Materials on Commerce and Industry] 3 (1984), 69–75; Sheng Mujie, "Ye Kuichu chuanlue," [A Brief Biography of Ye Kuichu] *Shanghai wenshi ziliao xuanji* [Historical Materials on Shanghai] 60 (1988), 116–20; and Tao Shuimu, *Zhejiang shangbang yu Shanghai jingji jindaihua yanjiu* [Zhejiang Merchant Clique and Shanghai's Economic Modernization] (Shanghai: Sanlian shudian, 2000), 66–8.
32. Papers of the National Commercial Bank, Q268.1.754, held at *SM*.
33. Jing Benbai letter to Fan dated November 19, 1920, in *JDJYZJ* I: 629.
34. Zhongguo renmin yinhang Shanghaishi fenhang Jinrong yanjiushi, comp. *Jincheng yinhang shiliao* [Historical Material of Jincheng Bank] (Shanghai: Renmin chubanshe, 1983), 26. Wu was forced to resign as the founding supervisor of the Bank of China by Zhou in 1912. See Zhongguo yinhang zonghang and Zhongguo dier lishi dang'anguan, comp., *Zhongguo yinhang hangshi ziliao huibian* [Collected Materials on the History of the Bank of China] 3 vols. (Beijing: Dang'an chubanshe, 1991), 40–5.
35. On Fan as a stockholder of *l'Impartial*, see Wang Pang, "*Dagongbao* de zijin yu gufen biandong qingkuang," [Sources and Changes of *Dagongbao's Capital*] in wwwpeople.com.cn/gb/shizeng/252/8397/8400/20020607/747102.
36. On Wu's association with Fan, see entry dated December 12, 1926, in "*Bian Baimei riji*" [Diary of Bian Baimei] held at the library of the Tianjin Political Consultation Conference; and He Lian, *He Lian huiyilu* [Memoirs of Franklin Ho] (Beijing: Zhongguo wenshi chubanshe, 1988), 89.
37. "*Bian Baimei riji*," entries dated April 9 and 10, 1920.
38. The Zhou and Sun families, relatives by marriage, were then supporters of Yuan Shikai. See Bao Peizhi, "Shouzhou Sunjia yu Zhongfu yinhang," [The Sun Family of Shouzhou and the Zhongfu Bank] *TJWSZL* 35 (1986),

137–51. On Bian's relationship with Zhou and Sun, see Brett Sheehan, *Trust in Troubled Times* (Cambridge, MA: Harvard University Press, 2003), 35.
39. Entries dated September 16, 1922, September 22, 1922, November 9, 1922, November 12, 1927, and December 27, 1927 in "*Bian Baimei riji*."
40. Zhongguo renminyinhang Shanghaishi fenhang jinrong yanjiushuo, comp. (1990), 128. The banks and bankers, of course, form their own overlapping and competing networks. See, for example, Brett Sheehan, "Myth and Reality in Chinese Financial Cliques in 1936," *Enterprise and Society* 6.3 (2005), 452–91.
41. Multiplexity tracks the exchange of multiple contents (material and immaterial resources), while transitivity refers to how a link "initiated by an actor passes through other actors," directly or indirectly. See Todeva (2006), 42–3.
42. Li Yuanhong (1864–1928), successor of Yuan Shikai as president of the republic, held 400 shares of the company under his family corporation *Dadetang*. See Jiuda's 1941 shareholder register.
43. On the theory of "structural hole," that is, intermediaries connecting networks by matching members' needs and interests, see Ronald Burt, *Structural Holes* (Cambridge, MA: Harvard University Press, 1992), passim.
44. See Chen Fan, ed., *Liang Shiyi shiliaoji* [Collected Historical Materials on Liang Shiyi] (Beijing: Zhongguo wenshi chubanshe, 1991), 350, 364.
45. Liang Shiyi letter to Fan Xudong dated May 24, 1918 with regards to Liang Qichao in *JD*.
46. Tang Peisong, son of Tang Shouqian held certificate number 81 with ten shares. On the Zhejiang network, see Zhejiangsheng Xinhai gemingshi yanjiuhui and Zhejiangsheng tushuguan, comp., *Xinhai geming Zhejiang shiliao xuanji* [Selected Historical Materials on the 1911 Revolution in Zhejiang] (Hangzhou: Zhejiang renmin chubanshe, 1981), passim.
47. Zhang Yufa (2002 ed.), passim.; and Liu Jingquan, *Beijing minguo zhengfu yihui zhengzhi yanjiu* [Parliamentary Politics in Republican Beijing] (Tianjin: Tianjin jiaoyu chubanshe, 2006), passim.
48. See Rao Huaimin ed., *Liu Kuiyi ji* [Collected Writings of Liu Kuiyi] (n.p.: Huazhong shifan daxue chubanshe, 1991), 148–9; Zhang Yufa (2002 ed.), 87–9. However, Liu's son insisted that Liu had served Yuan Shikai as a matter of strategy with the approval of Song Jiaoren and Huang Xing. See Liu An'nai, "Yi xianfu Liu Kuiyi shengping de pianduan yishi," [Bits of Memory on My Father Liu Kuiyi] *Xiangtan wenshi zilao* [Historical Materials of Xiangtan] 1 (1983), 17–29.
49. On Han, see Jiuda shareholder register (1941), and Yao Zhongyan and Tong Weidong, et al., "Da shiyejia Han Yongqing," [Han Yongqing the Industrialist] *Wuhan wenshi ziliao* [Historical Materials of Wuhan] 34 (1988), 175–83; and "Guangban cishan shiye de jushang Han Yongqing," [Philanthropist and Merchant Han Yongqing] in Wuhanshi zhengxie wenshiziliao weiyuanhui, ed., *Wuhan renwu xuanlu* [Selected Biographies of Wuhan Notables] (Wuhan: Wuhanshi wenshi ziliao faxing fuwubu, 1988), 332–4. His compradore past came back to haunt him in 1929 when the Nationalist government declared him a counterrevolutionary and forced him to flee Nanjing.

50. See, for example, Xu Foshu's letter to Fan Xudong dated March 16, 1919, in *JD*.
51. On Jing's involvement in politics and the National Commercial Bank, see Jing Benbai (1935), 5; Watanabe Atushi, "Shingai kakumeiki ni okeru ensei kaikaku undō," [On the Salt Reform Movement during the 1911 Revolution] *Kumamoto daigaku kyōiku gakubu kiyō* 27 (1978), 79. On Ye and Jiang, see Zhongguo renminyinhang Shanghaishi fenhang jinrong yanjiushuo, comp., *Shanghai shangye chuxu yinhang shiliao* [Historical Materials of the Shanghai Commercial Savings Bank] (Shanghai: Renmin chubanshe, 1990), 87; and *Zhongguo yinhang hangshi* bianji weiyuanhui, comp., *Zhongguo yinhang hangshi* [A History of Bank of China] (Beijing: Zhongguo jinrong chubanshe, 1995), 88. For Zhejiang politics, see Mary Backus Rankin, *Elite Activism and Political Transformation in China* (Stanford, CA: Stanford University Press, 1986), 248–98.
52. Xue Xianzhi, "Jiuda yanye gongsi shimo," [A History of Jiuda Salt Industries] *Tanggu wenshi ziliao* [Historical and literary materials of Tanggu] 3 (1992), 70. As the company expanded, more political figures became shareholders, including Cao Kun (governor of Hebei, 1916; president 1923–24), Chen Guangyuan, Wang Chengbin, Yang Jinglin, Li Yanqing, and other warlords of the period. See 1941 Jiuda shareholder register; Zhang Pengyuan, *Lixianpai yu Xinhai geming* [The Constitutionalists and the 1911 Revolution] (Taibei: Academia Sinica, 1969), 58; and Huang Yiping et al., *Beiyang zhengfu shiqi jingji* [The Chinese Economy during the Northern Warlord Period] (Shanghai: Shanghai shehui kexueyuan, 1995), 341.
53. Adshead attributed the delay to capital and technological difficulties. See Adshead (1991), 170.
54. English translation of Jiuda's first petition in *CLHB* 680.21.456. Foreign salt smuggling was reported in the lower Yangtze and Yunnan. See, respectively, Zhang Maojiong et al., eds., *Qing yanfazhi* [Salt Codes of the Qing Dynasty] (Beijing: Yanwushu, 1920), *juan* 86 and 87, and *The Times* (London) August 16, 1902.
55. Petition dated July 15, 1914. See *JDJYZJ* 1: 6–7, and Ben Bai, "Zhen bu ke jie" [Really Incomprehensible] repr. *YZCK* 2 (1931), 13–17. There is no evidence that the company added magnesium carbonate to aid the flow of salt, as Morton Salt Company began doing in 1911. See James Ballowe, *Salt and Trees* (Dekalb, IL: Northern Illinois University Press, 2009), 172.
56. Letter to the groups consortium dated April 30, 1914 in Ding An (1922), 2: 112a–b.
57. *JDJYZJ* I: 7. Dane was on home leave from June 9, 1914, until September 14, 1914. See Ding An (1922), 2: 69a and 2: 93a.
58. Presidential Special order on Salt Production reproduced in Zhongguo dier lishi dang'an'guan comp. (1991) 2: 1386–8; Jing Benbai (1935), 9; and Ding An (1922), 2: 137b–138a.
59. Supplemental regulations on the Special Order on Salt Production, repr. in *ZGYZSL* IV: 62–5.

60. For detailed descriptions of Jiuda's production process, see *JDJYZJ* 1: 128–31; and Lu Diping et al., eds., *Tianjin gongshangye* [Commerce and Industry of Tianjin] (Tianjin: Municipal Social Bureau, 1930), chap. 4, 4–5.
61. *ZGJDYWSZL* 1: 424–25; *JDJYZJ* 1:8–10.
62. *JDJYZJ* 1: 10.
63. On the rivalry and acrimony between Zhou and Zhang, see Zhou Zhijun, "Yuan Shikai dizhi huodong yu Yue-Yuanxi zhizheng," [Guangdong and Anhui Factionalism in Yuan Shikai's Restoration] *Wenshi ziliao xuanji* [Selected Historical Materials] 13 (1960), 89–93. See also discussion below on the transitive liability of networking.
64. *CLHB* 680.7.1077, Jing's petition to the Salt Administration dated September 22, 1915; Lauru's recommendation dated November 7, 1915; Gong's endorsement the next day *CLHB* 680.21.434; and Jing Benbai (1935), 9.
65. *CLHB* 680.7.1077, order issued March 4, 1916. On the issue of treaty ports, see Zhang Hongxiang, *Jindai Zhongguo tongshang kou'an yu zujie* [Treaty Ports and Foreign Concessions in Modern China] (Tianjin: Renmin chubanshe, 1993), Appendix 3, 4, and 6.
66. On this category of trade ports, see Zhang Jian, "Wanqing zikai shangbu shulun," [Voluntarily Opened Trade Ports in Late Qing] *JDSYJ* 83.5 (1994), 73–88.
67. Zhang Hongxiang (1993), appendix 3, 4, and 6; Ben Bai (1935), 10–11.
68. *JDJYZJ* I: 369–71.
69. See Li Dongjun, *Zhongguo sixue bainianji* [A Century of Private Education in China] (Tianjin: Nankai University Press, 2004), 275. Xudong visited Yan in person, see entry dated November 24, 1915 in *Yan Xiu riji bianji weiyuanhui*, ed., *Yan Xiu riji* [Diary of Yan Xiu], 4 vols. (Tianjin: Nankai University Press, 2001), IV: 2001. Fan Yuanlian had worked under Yan at the Board of Education in 1909.
70. *JDJYZJ* I: 10–11. Li Sihao and Jing graduated in the same class of the Zhejiang provincial examination. We have misdated the letters as 1915. The above account suggests that they should be dated as late as March 1916. This and other services by Li for Jiuda would cost the company 3,000 to 5,000 yuan in 1927, although he asked for a loan of 10,000. See Jing Benbai letter to Fan Xudong dated May 28, 1927 in *JD* 700260.
71. On Zhou's fall out with Yuan, see Zhou Zhijun (1960), 89–93.
72. Zhang Tongyi (1987) 22; and Ge Peilin, "Yang Du yu Tianjin," [Yang Du and Tianjin] *Tianjin shizhi* [Gazetteers of Tianjin] 1 (1998), 32.
73. *JDJYZJ* 2: 747–49.
74. Report to Changlu Inspectorate dated January 17, 1917 in *JD* 100919.
75. Systemic run of profit data for Changlu revenue farmers during the republican period is not available. Four undated ledgers from the Shijitang Li household operating from late Qing to 1928 the Wuqiang, Wuyi, Gu'an, and Anping monopolies suggest an after-tax net profit ranging from 1.138 (Anping) to 1.352 (Gu'an) *yuan* per *dan*. See "Tianjin Shijitang Lishi jiazhu wenshu," [Documents from the Shijitang Li Family of Tianjin], personal collection.

76. Jiuda petition to the Salt Administration dated June 29, 1916 in CLHB 680.7.1077 and brief to the Salt Administration dated December 24, 1916 in JD 700144–5.
77. Test conducted and wastage rate confirmed CLHB 680.7.1077, dated July 22, 1916; and JDJYZJ II: 750.
78. ZGJDYWSZL 1: 425–26; and Jing Benbai (1935), 9.
79. Salt Administration denial dated April 25, 1917 in CLHB 680.7.1078.
80. Reduced to 20 percent in 1921, and 10 by 1926. However, wastage allowance (luhao) for crude salt remained at 30 percent, putting refined salt at a severe disadvantage. See Zaiseibu Zeimushi enmuka, comp., Enmu shiryō [Materials on salt] (Shinkyō: Zaiseibu, 1935), 318.

3 Breaking into the Market

1. Ministry of Agriculture and Commerce order #800 on presidential citation of the company dated May 13, 1923, in CLHB 680.22.513.
2. On the distinction between bribery, defined as a transaction with an explicit quid pro quo, and lobbying with no results guaranteed, and rent extraction (zero-sum game) as opposed to rent creation (win-win game), see Fred McChesney, *Money for Nothing* (Cambridge, MA: Harvard University Press, 1997), 34–7.
3. Petition filed February 7, 1917; Changlu denial dated May 31, 1917, in CLHB 680.21.436; and Salt Administration denial dated June 13, 1917; company appeal dated July 6, 1917, claiming that the list was compiled from official records in CLHB 680.7.1080.
4. N. Ley, Changlu district inspector to chief inspectorate dated February 28, 1919, in CLHB 680.21.434.
5. Memo from Fujian and Yunnan commissioners dated October 13, 1916 and November 29, 1916, respectively, in CLHB 680.7.1079.
6. The Li and the Fang families, both from Ningbo and active in Shanghai as industrialists and the sugar trade, were also related by marriage. Together, members of the two families held 115 shares in the company by 1919.
7. Jing Benbai's letter to Fan dated October 28, 1916, and Fang Ji'nin's letter to Fan dated November 23, 1916, in JDJZJ I: 163–4.
8. Lianghuai merchant rejection in 1916 in JD 700152–54; Jing Benbai's letter to Fan Xudong dated October 23, 1916 in JD 700148–49 and Zhang's letters to Fan in 1916 in JDJYZJ I: 160–1.
9. Gong Shicai's letter to Fan dated 15th of unknown month, 1919 in JDJYZY I: 35–36.
10. Gong Shicai's letter to Fan dated April 23, 1919 in JD.
11. Contract between Jiuda and Wu Jinyin dated July 15, 1922 in JD 800505 and JDJYZJ I: 175–6.
12. Yue Ruizhou, commander of the Eastern Jiangxi garrison, became an early shareholder of the company (certificate 171, ten shares). Jinyin also served as the liaison person with Cao Kun for the company. See later.

13. Jérôme Barthélemy, "Agency and Institutional Influences on Franchising Decisions," *Journal of Business Venturing* 26.1 (2011), 93–103.
14. Zhang's letter to Fan dated January 22, 1916, in *JD*.
15. He Jichun, manager of Jiuda Changsha office, to Fan dated June 8, 1935. In other cases, the company sued former partners based on these contracts. For examples of such contracts, see Jiujiang's Yongchanghe in 1925 and Xinfa in 1934 in *JD*.
16. Jing Benbai's petition to the Changlu Salt Administration dated March 24, 1917; approval by Li Sihao dated April 23, 1917 in *CLHB* 680.7.1078.
17. Changlu guild petition dated August 12, 1916; Gongyi petitions dated August 13, 1916, September 5, 1916, September 17, 1916 over the starting date of the two-month notification period; Jiuda's rebuttal dated August 25, 1916; and Salt Administration order #342 in *CLHB* 680.7.1079.
18. Jing (1935), 14–15.
19. Lianghuai commissioner on behalf of Qingjichang dated March 4, 1917, and denial by the Salt Administration dated March 9, 1917 in *ZGJDYWSZL* I: 283–7.
20. An island on the Yangzi off the western wall of Nanjing city, Xiaguan was declared a treaty port by the Treaty of Tianjin in 1858. See Tang Wenqi, "Nanjing kaibu shimo," [Opening of Nanjing Port] in Yu Ming ed., *Xiaguan kaibu yu Nanjing bainian* [Opening of Xiaguan Port and a Century of Nanjing] (Beijing: Fangzhi chubanshe, 1999), 10–15.
21. See revenue farmers' petitions and Songjiang suboffice order to Jiuda for relocation of its Suzhou outlet in April, 1919 in *JD* 400001–49.
22. Salt Administration order #520 dated May 28, 1919 in *CLHB* 680.26.822.
23. *JD* 400284–87.
24. Yihexiang's plaint in *JD* 400302–305, and the Salt Administration's brief in *JD* 400306–8.
25. On Jiuda's futile attempt to access the Beijing market in 1918, see *CLHB* 680.7.1081, and 1084; and the Beijing salt syndicate's lawsuit in 1919 against Tongtaide for unlawful marketing of Jiuda table salt in *CLHB* 680.7.1085.
26. Jiuda's brief in *JD* 400310–315.
27. *JD* 400322–330.
28. Administrative Court decision in *JDJYZJ* I: 418–9.
29. Decision dated August 21, 1920 in *JD* 400322–330.
30. Jing Benbai (1935), 27–30; and his letters to Fan dated November 19, 1921 and September 17, 1923 in *JDJYZJ* II: 628–9 and I: 422. Also known as the poor legal courtyard (*Pinzhengyuan*), the Administrative Court was staffed by "learned" men, but not experts in the law, and handled less than a dozen cases each year. See Chen Guyuan, *Shuangqingshi yuwen cungao xuanlu* [Selected Left-Over Writings from *Shuangqingshi*] (Taibei: n.p., 1965), 170–1. Only one of the four judges hearing the case had formal training in the law. See Huang Yuansheng ed., *Pingzhengyuan caijue luquan* [Collected Decisions of the Administrative Court] (Taibei: Wu'nan tushu, 2007), 455–6. Allegations of justice miscarried by this court were reported

in *The Leader* on September 20, 1921 as translated in *ZGJDYWSZL* I: 293-5.
31. Jing's letter dated August 24, 1920 in *JD* 700191-2.
32. Dated September 20, 1920, October 24, 1920, and November 13, 1920 in *CLHB* 680.8.620; and petition dated November 26, 1920 for continued operation in *JD* 400377.
33. Gamble's memo to Pan Fu dated October 21, 1920 in *ZGJDYWSZL* I: 290-1; and Jing (1935), 30. Sir Reginald succeeded Dane, his fishing partner, in 1918. See India Office Records, political, and secret annual files IOR/L/PS/11/135 held at the British Library.
34. See Pan Fu's reply to Cao Kun (Jiuda shareholder and military governor of Zhili, 1916-1923) dated September 26, 1920 promising that he would accommodate (*quyu tongrong*) Jiuda within the law; Fan Xudong's cable to Pan dated December 21, 1920 reporting that Ding was still balking in *JD* 800334 and 800290 respectively; and Jing Benbai (1935), 30.
35. *CLHB* 680.8.17.
36. Jiuda's request to the Administrative Court dated July 4, 1921 and the court's reply dated July 23, 1921 in *JD* 400383-385.
37. Administrative Court reply dated August 18, 1921 to the Salt Administration's query in *JD* 400391-2; and Salt Administration's order #1486 dated August 31, 1921 demanding the Changlu commissioner to report on how the shipping permits were authorized in *CLHB* 680.8.620; and Pan Fu's bureaucratic reply to Li Yuanhong's letter in *JD* 700401.
38. Changlu commissioner's report dated September 8, 1921 and Salt Administration orders #1814 dated September 1921 and 2257 dated November 25, 1921 in *JD* 400391-2 and 400407, respectively.
39. Jing Benbai's petitions in *CLHB* 680.8.620.
40. Strickland's memo to Pan Fu dated September 30, 1921 in *ZGJDYWSZL* I: 295-6. See Jiuda's petition dated June 6, 1916 on the use of wood crates, and Esson Gale's approval of a seventeen-ton refined salt shipment to Hankou in cotton gunny bags (as used in soda ash shipments by foreign companies) dated March 2, 1917 in *CLHB* 680.7.1077 and 680.7.1078.
41. Zhong Shiming's reply to Strickland's memo dated November 4, 1921 in *ZGJDYWSZL* I: 296. Clause 4 of the Administrative Court Law promulgated on July 21, 1914 specified that its decisions could not be appealed. See Cai Hongyuan ed., *Min'guo fagui jicheng* [Compendium of Republican Statues], 100 vols. (Hefei: Huangshan shushe, 1999), XXXII: 28-33.
42. Jing Benbai's letters to Fan Xudong dated November 18 and 19, 1921 in *JD* 700406-411.
43. See Jing's letter to Fan Xudong dated December 24, 1921 in *JD* 700414-7 on his negotiations with Zhong, and Changlu Commissioner's approval of the product dated February 2, 1922 in *JD* 400426.
44. Liang's letter dated February 7, 1922 and Sun's reply in *JD* 700420.
45. Salt Administration order #600 in March, 1922 in *YZZZ* 34 (1922), *wendu*, 13-15. However, defining grain size remained a contentious issue. The Salt Administration dispatch 1227/30 dated February 18, 1930 authorized use of gunny bags in shipping both "refined large grain salt" and

"refined crystal salt," but the assistant district inspectors still found the product "not much different from the 'refined powered salt' to be exported by paper packets." See memo of Changlu District Inspectorate dated December 14, 1930. Indeed, refined powdered salt must still be shipped in paper packaging only. See *CLHB* 680.21.412.
46. Jing Benbai's petition in November, 1918 in *JDJYZJ* I: 144. Changlu Commissioner Ding Naiyang's approval of the purchase dated December 19, 1918 in *CLHB* 680.16.495. A personal letter from Chen Guangyuan, a company shareholder and military governor of Jiangxi from 1917 to 1922 requested a favorable decision. See Chen's letter to Ding in *JD* 800325.
47. Application dated March 3, 1919; denial dated March 22, 1919; and approval by order #791 dated June 5, 1919 in *CLHB* 680.7.1083–5.
48. However, Jing also wrote that the shipment never made it to the Soviet Union. See Jing Benbai (1935), 44.
49. Jiuda's petition dated August 20, 1920 and Salt Administration's denial dated September 4, 1920, both in *CLHB* 680.7.1089.
50. Another application to raise the limit by 300,000 *dan* was made on October 20, 1921, supported by Cao Kun's personal cables to the Salt Administration on October 28 and 31, 1921. See *CLHB* 680.7.1093.
51. Jing Benbai (1935), 37–43.
52. Letter from the Administrative Court to Yongli in *JD* 700485 and Salt Administration order #1347 dated October 26, 1923 notifying Jiuda of the Administrative Court's decision in *CLHB* 680.78.1094. See also Jing Benbai's letter to Fan Xudong dated September 17, 1923 on his discussions with secretaries at the Salt Administration and possible access (*chulu*) through Liu Songsheng, Jiuda's legal counsel (and shareholder) to the Administrative Court. Receiving four months' pay was considered a good year for the court. See respectively *JD* 700465 and Chen Guyuan (1965), 170–1.
53. *YZZZ* 37 (1923), *jishi*, 1–5; and Jing Benbai (1935), 35–6.
54. Huang Guoxi's (Jiuda's Hankou manager) letter to Fan Xudong dated May 17, 1923 in *JD* 700432; and Hubei provincial assembly's letter of transmittal to the governor, *YZZZ* 38 (1923), *wendu*, 1–2.
55. Tang Binru's letter sent at Li's (president from June 10, 1922 to June 13, 1923) instruction dated September 10, 1922 in *JDJYZJ* II: 556–7.
56. R. D. Wolcott's letter dated December 14, 1923 and Cao's letters to Pan Fu on behalf of the company cited above. Cao, military governor of Hebei from 1916 to 1923 and president from 1923 to 1924, was recruited as a shareholder through Chen Guangyuan. See Chen's letter inviting Cao to join Jiuda as a shareholder and introduction of Wu Jinyin. Dated January 13 of unknown year (probably 1919) in *JD* 800326.
57. Huang Guoxi's letter to Fan Xudong dated July 16, 1923 in *JD* 700447.
58. Privately, Huang found twenty bags acceptable, although thirty was better. See Huang Guoxi's letters to Fan Xudong dated June 21, 1923 and August 17, 1923 in *JD* 700442 and 700456.
59. Hubei Provincial Salt Bureau to Salt Administration in *JD* 400838–9.
60. On Ye, Qian, and Zhang, see Jing's letters to Fan Xudong dated September 17, 1923 and October 1, 1923 in *JD* 700464 and 700469, respectively.

61. Orders issued October 10, 1923 and December 24, 1924 in *JD* 700495 and 400485 respectively.
62. High officials in the Salt Administration such as Fu Dingyi (146 shares) and Zhu Youping (transport and market bureau, a graduate of Oxford University, 30 shares) also held Yongli shares.
63. On the theory of strong ties for survival and weak ties for growth, see John Watson, "Modeling the Relationship between Networking and Firm Performance," *Journal of Business Venturing* 22.6 (2007), 852–74.
64. While stationed at Yichang, Wolcott accepted silverware and a squirrel skin from Jiuda, a company "in a way indebted" to him. He also reported "stories that Gale was deeply mixed up in the big wastage allowance scandals of the Yangtze Valley, and is a welathy (sic) man." See respectively his letters dated December 14, 1923 and March 25, 1928 in R. D. Wolcott letters. Li Yanqing and Wang Lanting, Cao Kun's key aides, each received 500 shares of Jiuda and Yongli. See letters dated December 20, 1922 and January 13, 1923 in *JD* 700429 and 800297. On loans to the Administrative Court, see *JD* 700465, 700478 and 700485. Zhang Guogan (1876–1959), head of the court from 1920 to 1922, was a shareholder of both Jiuda and Yongli under various nominees. He was succeeded at the court by Wang Daxie, another shareholder.
65. On accessing a commander of Hubei's revenue guards, see Huang Guoxi's letter to Fan Xudong dated October 12, 1923 in *JD* 700473.
66. Joseph A. Schumpeter, *Capitalism, Socialism and Democracy* (London: Routledge, 1994), 82–3.

4 The Price of Success

1. I owe this insight into the tension between creative destruction and conservative restrictive practices in capitalism to Simon Glezos, "Creative Destruction versus Restrictive Practices: Deleuze, Schumpeter and Capitalism's Uneasy Relationship with Technical Innovation," http://www.ctheory.net/articles.aspx?id=648, accessed April 19, 2013.
2. *ZGYZSL* III: 129. Nie Qijie, a Yongli founding shareholder (and elder brother to Nie Qiwei, see below) urged cooperation between Jiuda and Lequn but was rejected by Xu Jingren, a major Lianghuai producer. See *YLHZ* I: 22–3. Lequn's inventory was liquidated as crude salt in 1933. See Caizhengbu Yanwu jihesuo, comp., *Min'guo ershinian Yanwu jihesuo nianbao* [Annual Report of the Revenue Inspectorate for 1933] (Nanjing: Caizhengbu, 1934), I: 18.
3. Registration application dated June 14, 1921, ruling issued August 26, 1921, and rejection dated October 1, 1922 in *CLHB* 680.8.719. The company finally received approval on June 5, 1922.
4. With investment from Yuan Shikai's family. See Wang Qian'gan, "Lun Anhui Shouxian Sun Jia'nai jaizu dui Zhongguo jindai jingji fazhan de gongxian," [Contributions to Modern China's Economic Development by Sun Jia'nai's Family from Shouzhou, Anhui] *Min'guo dang'an* [Republican Archives] 2 (2004), 106–13.

5. Shanghaishi liangshiju, et al., comp., *Zhongguo jindai mianfen gongyeshi* [Flour Industry in Modern China] (Beijing: Zhonghua shuju, 1987), 190–209. See also Zhongfu's papers held at *SM*.
6. On the bank, see Bao Peizhi, *TJWSZL* 35 (1986), 137–51. On Tongyi's registration and shareholders, see *ZGYZSL* II: 112–3, and the company's files in *SM* and *AS*.
7. On the company, see *JD* 701249, 17–23–072–32–000–021 in *AS*, and Bankers' Cooperative Credit Service reports, Q275 held at *SM*. The Zhou household of Nanxun held extensive interests in salt, silk, and native banks. Zhou himself was also active in the Zhejiang provincial railway movement.
8. Letter from Chen Canglai, Jiuda's Shanghai manager, to Fan Xudong dated May 12, 1929 in *JD* 701247–8. On the washing process, still deemed unhygienic, see Chen Canglai, *Zhongguo yanye* [China's Salt Industry] (Shanghai: Shangwu yinshuguan, 1928), 53.
9. Petition of Mingsheng Refined Salt Company in *Liangzhe yanwu guanliju* [Papers of the Liangzhe Salt Administration], L057–14–91 and L057–14–92 held at the Zhejiang Provincial Archives, Hangzhou.
10. Including Hengyuan Salt Refinery with a proposed capital of 1 million *yuan* and an annual production capacity of 150,000 *dan*. See *CLHB* 680.8.721.
11. This pattern differs from studies that identified the golden age of the republican economy with either before or after 1920–1926. See, respectively, Shi Bolin, *Qifeng kuyu zhong de minguo jingji* [The Precarious Republican Economy] (Zhengzhou: Henan renmin chunbanshe, 1993), 223; Fang Qingqiu, et al., *Minguo shehui jingjishi* [Social and Economic History of the Republican Period] (Beijing: Zhongguo jingji chubanshe, 1991), passim.; and Marie-Claire Bergère, *The Golden Age of the Chinese Bourgeoisie, 1911–1937*, transl. Janet Lloyd (Cambridge: Cambridge University Press, 1989), passim.
12. By 1935, there were twenty-one registered salt refineries, although Du Xuncheng listed only twelve. See his *Minzu ziben zhuyi yu jiu Zhongguo zhengfu* [National Bourgeois Capitalism and Old China's Government] (Shanghai: Shanghai shehui kexueyuan chubanshe, 1991), 401–2. For the registered refineries and forty failed applications, see Zaiseibu Zeimushi enmuka, comp. (1935), 315–6; and *CLHB* 680.8.595; 680.8.721–3; 811–6; 680.9.75; 680.16.569; 680.21.455; and 680.22.873.
13. Minutes of shareholders' meeting, March 21, 1920.
14. However, as several franchisees who converted their security deposits into company shares had redeemed their shares, 90,030 *yuan* of the 2.1 million was not paid-up. See *JD* 200948.
15. On these companies, see *JDJYZJ* I, chap. 9; and Song Meiyun and Zhou Licheng, eds., *Chuanwang Dong Haoyun zai Tianjin* [Dong Haoyun the Shipping Magnate in Tianjin] (Tianjin: Renmin chubanshe, 2008), passim.
16. Ben Bai, "Yongyu yanye gongsi chuangli shi," [A History of Yongyu Salt Industries] *YMZK* (1935b), 44–5.
17. On the negotiations, see Zhongyang yanjiuyuan Jindaishi yanjiusuo, ed., *Zhongri guanxi shiliao* [Historical Materials on Sino-Japanese Relations] (Taibei: Academia Sinica, 1995), passim.

18. See the company's charter and shareholders' list in 17-23-0-75-37-000-003 held at *AS*; Qingdaoshi shehuiju comp., *Qingdaoshi yanye diaocha* [Survey of Qingdao's Salt Industry] (Qingdao: Qingdaoshi shehuiju, 1933), 6-10; Ben Bai (1935b), 45; and Shoji Tsutomu, comp., *Nihon soda kogyōshi* [A History of Japan's Soda Ash Industry] (Tokyo: Soda sarashiko dogyōkai, 1938), 147-52.
19. On the contract between Yongyu and Jincheng, see papers of the Yongyu Salt Industries, Ltd., and the bank's papers J211.4325 held at *TM*.
20. Letters from Jiuhe to Hengfengtang, *JD* 701216 and 701232.
21. Minutes of Jiuhe Board Meeting dated June 15, 1929 in *JD* 701249-250.
22. Including a rebate ranging from ten to thirty cents per *dan*, and packing three extra *jin* to each bag free of charge. See letters to Fan Xudong dated January 24, 1924, June 20, 1924, July 4, 1924, and September 18, 1925 in *JD* 200654, 700501-3, 700532 and 700563-4. The Hongkong Bank, for instance, often refused to handle local native bank orders. See *HSBC* SHG II, 296.
23. Letter dated January 29, 1924 in *JD* 700500.
24. Xu Zhaozhou's letter to Fan in *JD* 700530.
25. Fan's letter to Miao Qiujie in 1925. See *JD* 800312.
26. Cable to Tan Yankai dated August 11, 1920 in *JD* 800288-9.
27. *JD* 400702, 400706-9, 400710, 400715-6, 400724-8.
28. Franchisee letter dated December 14, 1923 in *JD* 400712; and company petition to the Salt Administration dated December 27, 1923 in *CLHB* 680.7.1094.
29. *JD* 700432, 700566.
30. A total of 560 shares through various nominees. His son Guoyin once served as a company director. See Qi Xiemin, "Wosuo zhidao de Chen Guangyuan," [The Chen Guangyuan That I Knew] *TJWSZL* 36 (1986), 120-31.
31. *JD* 700394; *Yishibao* [*The Social Welfare*]November 5, 1922.
32. "Military Interference in Salt Gabelle: A Critical Situation," *The China Weekly Review* XXIX: 15 (1924), 4. See also Fan Xudong's cable dated August 26, 1922 to Wu Jinbiao asking him to solicit from Cao Kun a few kind words on behalf of Jiuda in *JD* 800293.
33. *JD* 700572-6.
34. Letters from Jiujiang franchise to Jiuda head office dated February 23, 1925 and March 21, 1925; and Fan's letter to Lu dated October 14, 1925 in *JD* 800310. Lu held eleven shares (certificate #268).
35. Other alternative approaches included a brother of the chief secretary at the provincial treasury; or Qi Zongtang, Fang's in-law. Letters from Jiujiang franchise to Fan dated March 21, 1925 and March 25, 1925 in *JD* 700568-571.
36. Xu (1873-1964) served briefly as governor of Anhui in 1924, premier, and concurrently finance minister in 1925.
37. *JD* 700197-201. Finally defeated in battle, Fang was replaced by Deng Ruzhuo in 1926. With letters from Ni Daolang, another Jiuda shareholder whose father, Sichong (1868-1924, who controlled neighboring Anhui for many years) had sponsored Deng's appointment, Jiangxi's refined salt

market was restored. See letters from Jiujiang franchise to Fan dated March 6, 1925 in *JD* 700571b; and Ni's letters to Deng's nephew and aide dated April 5, 1925 in *JD* 800328–9.
38. See, for examples, Cao's cable to the Salt Administration supporting the company's application for an increased production quota dated October 28, 1921 and October 31, 1921.
39. See correspondence between Fan and Zhang Hu dated August 19–20, 1925 in *JD* 800429–34.
40. Li, a close aide to Cao Kun, also received a gift of 5,000 yuan in Yongli shares. See Zhou Xueting letter to Li dated January 13, 1923 in *JD 800297*.
41. Bureau of Warfare Relief Order dated June 12, 1925 and Jiuda's reply dated June 15, 1925, and meeting with the associate chief inspector-general dated August 19, 1925. As negotiations with local authorities went nowhere, Jiuda approached the Fengtian warlords through Yun Baohui (a shareholder), then secretary to Bao Guiqing, governor of Helongjiang. With the return of the shares to their rightful owners in 1926, the company was compensated with 80,000 yuan in provincial bonds of dubious market value.
42. Wu's order issued March 28, 1926 in *JD* 700193; and Chen's orders in *JD* 700233–249. Tongyi, Chen argued, had paid his Shandong counterpart 200,000 *yuan* and still made another 200,000 *yuan* in profits for the year.
43. *CLHB* 680.21.435.
44. Letter from Kang Youwei to Yang Yunshi, Wu's aide, dated July 25, 1926 in JD 700205.
45. Petition to the Salt Administration in September, 1926 in *JD* 701364, 700206.
46. Fan's letter to E. W. Mead dated September 25, 1926 in *JD* 700213, with further arguments in 700206, 700224, 700241–4, and 701368.
47. Fan Xudong's correspondences with Fu Dingyi, the Revenue Inspectorate, and the Salt Administration in *JD* 700217 and 800313.
48. Jing Benbai's letter to Fan Xudong dated September 7, 1926.
49. Jiuda's branch offices in Shashi, Wuhu, and Hunan were dissolved and reorganized in 1927 and 1928. See Jiuda head office memos dated March 9, 1927, August 22, 1927, June 28, 1928 in *JD* 200950–1.
50. *ZGJDYWSZL* I: 409–410.
51. Chen Xinwen erred in identifying this bureau with Zhu Yupu and that Jiuda did not pay the bureau to resolve the matter. See Chen Xinwen (2002), 28.
52. Letter to L.C. Wolcott dated July 16, 1927 in R. D. Wolcott Letters.
53. Jing Benbai letter to Fan Xudong dated February 8, 1928 in *JD* 101376.
54. Entry dated November 12, 1927 "*Bian Baimei riji*"; in Zhou Zuomin's letter to Fan Xudong dated 28th (January 28, 1928, but could be as early as December 1927), and Fan Gaoping (one time Jiuda's Nanjing manager) as a classmate to Sun Shiwei, the new provincial governor, in *JD*.
55. Jing Benbai's petition to the Changlu Salt Commissioner dated January 27, 1928 in *CLHB* 680.21.444; and denial by the inspectorate dated February 15, 1928.

56. Jing Benbai letter to Fan Xudong dated February 25, 1928 in *JD*; and Chief Inspectorate to Changlu District inspector dated April 4, 1928 in *CLHB* 680.21.444.
57. Changlu Salt Commissioner to Zhu Yupu dated December 24, 1927 in *CLHB* 680.21.444; and *JDJYZJ* I: 604–20.
58. The deflated 0.709 yuan per *dan* profit rate recorded in Jiuda's 1920 Annual Report is the result of the company's heavy capital outlay that year, including electrification of the refineries and installation of pumps for the refurbished evaporation ponds. An extraordinary provision of 110,000 yuan for bonus shares issued to existing shareholders further reduced the rate of profit. See Jiuda Annual Report for 1920.
59. Hu Jinyin's letter dated July 24, 1921 and Cao Kun's note acknowledging receipt of unspecified gifts from Jiuda dated May 27, 1922 in *JD* 800332–3.
60. ReardonAnderson is imprecise in stating that Yongli began "regular production" in 1924. See his *The Study of Change* (New York: Cambridge University Press, 1991), 165. Plant commissioning did begin on January 10, 1924, and soda ash of unmarketable quality started flowing from March 8, 1924 until June 26, 1924. Production then resumed on April 5, 1925 and ceased again on November 1, 1925. Yongli's plant commissioning problems were not solved until June 27, 1926. See *YLZJ* I: 26–30.
61. See contract for a 600,000 yuan line of credit from Jincheng to Jiuda dated February 13, 1925, increasing to 750,000 *yuan* on September 25, 1926, secured on all its properties and inventory (including holdings of Hengfengtang). See J211.3601 and J211.4325 held at *TM*.
62. Memoirs of Yu Xiaoqiu held at *SM*.
63. "Report of the bank syndicate on Yongli," in Dec., 1934 in Papers of the Shanghai Commercial and Savings Bank, Q275.1.663, held at *SM*.
64. Minutes of shareholders' meeting, December 6, 1928.
65. Transcript of Fan's speech in the 14/15th shareholders meeting, 1928.
66. *ZGYZSL* I, Hubei, 2; Jiangxi, 2.
67. Ding Changqing, "Guomindang Nanjing zhengfu yu jiu yanshang," [The Nationalist Nanjing Government and the Traditional Salt Merchants] *YYSYJ* 1 (1988): 40–8.
68. Zhou Zuomin's letter to Fan requesting loan settlement dated November 25, 1929 in *JD* 700329; and Hu Zurong (Jiuda's chief accountant) reporting to Fan on National Commercial's refusal to renew another loan on February 26, 1930 in *JD* 700336–7.

5 Cartel as a Business Network

1. George W. Stocking and Myron W. Watkins, et. al. *Cartels in Action: Case Studies in International Business Diplomacy* (New York: Twentieth Century Fund, 1946), passim.
2. Margaret C. Levenstein and Valerie Y. Suslow, "What Determines Cartel Success?" *Journal of Economic Literature* XLIV (2006), 43–95, esp. 49–75.

3. On the debate over defining cost, see Phillip Areeda and Donald F. Turner, "Predatory Pricing and Related Practices under Section 2 of the Sherman Act," *Harvard Law Review* 88.4 (1975), 697–733. Economists and legal scholars still differ in how to measure cost and profit. See Peter Z. Grossman, ed., *How Cartels Endure and How They Fail: Studies of Industrial Collusion* (Cheltenham, U.K.: Edward Elgar Publishing, 2004), 2–7; Massimo Motta, *Competition Policy: Theory and Practice* (Cambridge: Cambridge University Press, 2004), 447–9; and Christopher R. Leslie, "Achieving Efficiency through Collusion: A Market Failure Defense to Horizontal Price-fixing," *California Law Review* 81.1 (1993), 243–92.
4. Wyatt Wells, *Antitrust and the Formation of the Postwar World* (New York: Columbia University Press, 2002), passim.
5. For cartels in the period, see, for examples, Guo Shihao and Sun Zhaolu, "Cong Qixin yanghui gongsi kan jiu Zhongguo shuiliye de longduan huodong," [Monopolistic Activities in Old China as Seen from the Case of Qi'xin Cement Co., Ltd.] *Jingji yanjiu* [Economic Studies] 9 (1960); Jennifer Zhang Ning, "Vertical Integration, Business Diversification, and Firm Architecture: The Case of the China Egg Produce Company in Shanghai, 1923–1950," *Enterprise and Society* 6.3 (2005), 419–51; and Chan Kai Yiu, *Business Expansion and Structural Change in Pre-war China: Liu Hongsheng and his enterprises, 1920–1937* (Hong Kong: Hong Kong University Press, 2006), passim.
6. Peter Z. Grossman (2004), 7.
7. Annual report of Tunway Enterprises for 1925 in *CLHB* 680.22.1383.
8. The idea to collaborate with Tongyi first came up in late 1924 when the Shashi market became accessible. See Xu Zhaozhou's letters to Fan Xudong dated October 16, 1924 and November 8, 1925 in *JD*.
9. Huang, a native of Jiangxi, began as a native banker at Wuyue where he befriended Feng Yuxiang and became the Christian general's business partner. His partnership also held Tongyi's Jiujiang franchise capitalized at 40,000 yuan. In 1927, he was elected chairman of the Hankou Chamber of Commerce. See *Wuhan renwu xuanlu* (1988), 353–4.
10. Xu Zhaozhou letter to Fan Xudong dated November 13, 1924.
11. Contracts between Hengfengtang and Tongqing in January 1926; between Datong and Jiuda in February 1926; and between Tongyi and Tongqing dated January 1, 1925 in *JD* 800516-9; 800520-2; and 800510-2 respectively.
12. *JDJYZJ* I: 555–6, Chen's order dated March 28, 1926.
13. *JD* 700194 and 700204. Zhang Tongyi erred in dating the organization to 1918. See Zhang Tongyi (1987), 22.
14. Undated manifesto in the Archives of the NSRB in *JY*.
15. Chen Xianchen's undated letter in *JY*.
16. *Dongfang shibao* [Eastern Times] August 20, 1926.
17. Revised charter of the NSRB, 1926 in *JY* 100014. The ministry's approval was contingent upon a change of name from trade union (*gonghui*) to *zonghui*, and removal of adjudication of disputes (*gongduan*) as a function of the association. See *YZZZ* 46 (1927), *jishi* 2.

18. Jiuda Hankou Office to Tianjin head office dated April 30, 1926 in *JD* 800526. Zhang Tongyi erroneously dated this association to 1918. See Zhang (1987), 157.
19. Letters from Xu Zhaozhou to Fan Xudong dated April 2, 1926 in *JD* 800526.
20. See, for examples, Pan Fu's letter to Fu Dingyi dated March 20, 1927, and Fu's reply dated March 24, 1927 in *JD*.
21. Xu Fuqi, "Huiyi huaxue gongyejia Fan Xudong," [Remembering Fan Xudong the Chemical Industrialist] *Jiangsu wenshi ziliao* [Historical materials of Jiangsu] 25 (1988), 27. Xu was Fan's nephew. Xinfu's last available ledger dated July 29, 1927 listed a total gross income of 535,920 yuan, against 205,619 in taxes, 208,000 paid to Jiuda, and 25,100 in special expenses (*tebie yongfei*). See *JD*.
22. Correspondence between Jiuda and Datong dated March 1, 1927.
23. Correspondence between Datong and Jiuda dated March 22, 1927, April 5, 1927, and May 3, 1927 in *JD*. Finally, the Refined Salt Association arranged a loan for 350,000 yuan.
24. See *JD* 100861, 700263, 700272, and Jiuda annual report for 1927. Dissolution did not occur until 1929.
25. Letter of Xiao Yegui to Li Zhoufu in *JDJYZJ* I: 172.
26. *JD* 700264, Shen Huakui letter to Fan Xudong dated November 8, 1929 in *JD* 700326.
27. *ZGYZSL* (1933), I: Hubei, 2. Xiao Yegui letter to Fan Xudong dated April 8, 1928 in *JD* 700319 and Jing Benbai letter to Fan in December 1931 in *JD* 101409. On the strike from April to September 1929 orchestrated by the local branch of the Nationalist party, see *JD* 200977–201009 and *Tianjin shuangzhou* [Tianjin Bi-weekly] 1.5 (1929), 6–8.
28. Xiao Yegui letter to Fan Xudong dated April 11, 1929 in *JD* 700320. Loan contract between the Hankou Refined Salt Association and the Salt Administration dated September 9, 1929 in *CLHB* 680.21.1001.
29. Letter from Liyuan to Jing Benbai dated January 22, 1930 in *JY*.
30. Zhang Shuyuan (manager of Jiuda Hunan office) letter to Fan Xudong dated February 14, 1932 in *JD*.
31. Letters from Jing Benbai to Fan Xudong dated February 3 and 14, 1930 in *JY*.
32. Letters from Jing Benbai to Fan Xudong dated February 14 and 19, 1930; and draft on the organization of cooperatives in *JY*.
33. See my "Market and Network Capitalism: Yongli Chemicals Co., Ltd. and Imperial Chemical Industries, 1917–1937," *Bulletin of the Institute of Modern History Academia Sinica* 49 (2005), 93–126.
34. Jiuda Annual Report for 1930, *JD* 701142–49; and Jing Benbai (1935), 51.
35. Letter from Hu Gengyu (Jiuda's chief accountant) to Fan Xudong dated March 8, 1930; and Xiao Yegui to Fan Xudong, June 29, 1932 in *JD*.
36. Memorandum signed by Jiuda, Wuhe, Yongyu and Liyuan, ca. February 1930, in *JY*.
37. Minutes of the Joint Office of Salt Refineries meetings held from February 28, 1930 to April 9, 1930 and July 28, 1930 in *JY*. On Hankou contract, see 200655; Jiujiang contract for 1930–1931, see *JD* 200466–7. Anqing

did not begin until August 6, 1931 because of a particularly contentious market.
38. Hankou chapter pooled sales agreement signed June 4, 1930 in *JY* 200655 and pact signed June 17, 1930 as reported in letter to Fan Xudong, *JD* 700343.
39. Resolution of the Joint Office dated July 28, 1930 in *JY*.
40. Jiuda letters to Yongyu dated November 23, 1930 and December 5, 1930; letter from Yongyu to Jiuda dated December 26, 1930 in *JD*.
41. Minutes of Joint Office meeting dated December 5, 1930 in *JY* 100359.
42. Or 26.7 percent (Jiuda and Yongyu combined) out of a total of 1,219,384 *dan* shipped by all the refineries.
43. Minutes of Joint Office meetings, May 14 to 25, 1931. Yuhua and Tongda finally relented and signed on May 31, 1931.
44. Minutes of Joint Office meetings dated May 19–20, 1931. This in effect reduced by nine percent every refinery's shipment.
45. Regulations on pooled sales adopted May 25, 1931.
46. *JY* 100434–438. The Joint Office papers contain no lawsuits over enforcement of these regulations, leaving open the question whether they could be enforced in a court of law.
47. By private arrangement, Jiuda accommodated Fuhai while Wuhe gave up one share of Hankou sales to Tongda in exchange for a 3,000 yuan payment. See minutes of Joint Office meeting dated May 24, 1931.
48. Jing Benbai letter to Fan Xudong dated July 26, 1932 in *JD* 700720.
49. *JD* 100611. The Hua'an Refined Salt Co., Ltd. was capitalized at 400,000 yuan, of which 320,000 came from shareholders of Hongyuan. See the company's shareholders list held at *AS*.
50. Wuhe's Shashi franchisee failed to pay the requisite local levies and a "bandit suppression" surcharge, and because of the collective guarantee, local authorities arrested a Jiuda employee. See minutes of Joint Office meeting dated May 17, 1932 and letter of Xiao Yegui to Fan Xudong dated June 3, 1932 in *JD* 800820.
51. Zeng Yangfeng, *Zhongguo yanzhengshi* [A History of Salt Administration in China] (Repr. Ed., Taibei: Shangwu yinshuguan, 1978), 37–8; and Li Han (1990), 54–100. By 1934, modern banks, led by the Bank of China, provided close to 30 million yuan of financing to the revenue farmers in the four Yangzi provinces. See Q275.1.1632 held at *SM*.
52. Jiuda's petition to the Changlu Commissioner dated June 27, 1932 in *CLHB* 680.7.1097; and *JD* 101424–7.
53. Fan Xudong letter to the Joint Office dated May 21, 1932, and Xiao Yegui quoting Fan in letters dated June 29, 1932, July 23, 1932, and August 29, 1932 in *JD* 100601, 100650, and 800823 respectively.
54. Chen Canglai letters to Fan Xudong dated July 13–19, 1932 in *JD* 700712, 700714.
55. Wang Xuhe letter to Fan Xudong dated July 21, 1932 in *JD* 100489, and Fan's reply dated July 25, 1932 in *JD* 100492.
56. Wang Xuhe letter to Fan Xudong dated July 26, 1932 in *JD* 100494–6.

57. Fan Xudong cables to Wang, Joint Office, and Jiuda branch managers dated July 29, 1932; and Xiao Baowen letter to Fan Xudong dated July 5, 1932 in *JD* 100585 and 100823.
58. *JD* 800910; and Yanwu jihesuo, comp., *Minguo ershiernian Yanwu jihesuo nianbao* [Annual Report of the Salt Revenue Inspectorate for 1933] (Nanjing: Yanwu jihesuo, 1934), 17–18.
59. Xiao Yegui letter citing Fan dated August 29, 1932 in *JD* 100601.
60. See "Yongli Jiuda guchang gongchangshi," [A History of Yongli and Jiuda Tanggu Plants], 71–2 in Tianjin municipal chemical industry bureau papers, X188.321 held at *TM*; Li Wencai, "Huailian wo de fuqin," [Remembering my father] *Xiangxi wenshi ziliao* 25–26 (1992), 159; and interview with Li Mingzhi, January 19, 2010.
61. Minutes of Jiuda shareholders' regular meetings dated March 26, 1922, April 20, 1924, and *JD* 200970. On the discussions, see Fan's undated letters to Yan Youpu reprinted in *Haiwang* 20.26 (1948), 404; 20.27 (1948), 426; and Wang Yi'nong, "Cong renzhi dao fazhi," [From Ad Hoc Decision Making to Formal Governance] in *Haiwang* 21.8 (1948), 113–18.
62. *JDJYZJ* I: 29–43.
63. "Ben tuanti xintiao," [Our Group's Guiding Principles] *Haiwang* 7.1 (1934), 25–9.
64. Chen Xinwen (2002), 117–18. Shared among the salt, soda ash, and caustic soda divisions, these expenses added 13,809 yuan to production cost in 1937. See "Yongli Jiuda guchang gongchangshi" [A History of Yongli and Jiuda Tanggu Plants], 30 in Tianjin municipal chemical industry bureau papers, X188.321 held at *TM*.
65. Fan Xudong's memos dated September 26, 1932 in *JD* 100621 and 700606. By 1934, Yongyu managed to reduce its cost from 2.605 yuan in 1931 to 1.704 yuan/dan (before interest and administrative charges which amounted to 0.337 yuan).
66. Letter of Xiao Yegui to Fan Xudong dated June 9, 1933 in *JD* 800956/958; and Fan's order dated June 18, 1933 in *JD* 800963.
67. As in the case of the Convention de l'Industrie de l'Azote for ammonium sulfate, "one of the big international cartels of the depression years." See Kim Coleman, *IG Farben and ICI, 1925–1953* (London: Palgrave, 2006), 38–42.
68. Report on Tongyi Salt Refinery, Ltd. dated December 28, 1934 in the Archives of the Bankers' Cooperative Credit Service, Ltd. (*Zhongguo jingxinsuo*) Q275.1.1991 held in *SM*. On Fan's strategy, see Li Tifu letter to Xiao Baowen dated September 18, 1933 and references to "sons of Hunan" in letters dated November 17, 1932, January 15, 1933 in *JD* 200822, 800870, and 800920.
69. Wells (2002), 9.
70. See shareholders registers of the companies held at *AS* and *SM*.

6 Relationship with the Nationalist State

1. See, for examples, the works of Lloyd Eastman, Parks Coble, and more recently, Zhang Lijie (2011).

2. Richard C. Bush, *The Politics of Cotton Textiles in Kuomintang China* (New York: Garland Press, 1982), passim; and Chen Zhongping (2011), passim.
3. See for example Adshead (1970), Julia Strauss (1998), and Yao Shundong, "Nanjing guomin zhengfu chuqi shiyan lifa yu Zhongguo yanwu jindaihua," [Alimental Salt Legislation by the Early Nanjing National Government and Modernization of China's Salt Administration] *YYSYJ* 1 (2007), 19–26.
4. Zhang Naiqi, *Zhongguo jingji xianshi jianghua* [Speeches on China's Present Economy] (Shanghai: Shenbao yuebaoshe, 1935), 22; Wang Fangzhong, "1927–1937 nianjian de Zhongguo yanwu yu yanfa gaige de liuchan" [China's Aborted Reform of Salt Administration and Law, 1927–1937], in Peng Zeyi and Wang Renyuan, eds., *Zhongguo yanyeshi guoji xueshu taolunhui lunwenji* [Proceedings of the International Symposium on the History of China's Salt Industry] (Chengdu: Sichuan renmin chubanshe, 1991), 159–81; Ding Changqing, "Guomindang Nanjing zhengfu yu jiuyanshang" [The Nationalist Nanjing Government and the Traditional Salt Merchants] *YYSYJ* 1 (1988), 41–48; and Zhang Lijie, "Tanxi Nanjing guomin zhengfu weineng shishi xinyanfa de yuanyin" [Why Did the Nanjing Government Fail to Implement the New Gabelle Law?] *YYSYJ* 3 (2007), 10–20.
5. *Tianjin jianchangzhi* bianxiu weiyuanhui, *Tianjin jianchangzhi* [A Gazetteer of Tianjin Soda Works] (Tianjin: Renmin chubanshe, 1992), 749; and Chen Xinwen (2002), 28.
6. David Kang, *Crony Capitalism* (Cambridge: Cambridge University Press, 2002), 3. See also Mark Granovetter, "The Social Construction of Corruption," in Victor Nee and Richard Swedberg, eds., *On Capitalism* (Stanford, CA: Stanford University Press, 2007), 152–174.
7. Proposal by Zhuang Songpu and Wei Tingsheng as drafted by Jing Benbai. See Jing (1935), 13. The National Guild of Condiment and Wine makers also made a similar proposal. However, the proposals from the Salt Administration and two leading revenue farmers made no mention of the issue. See Quanguo jingji huiyi mishuchu, comp., *Quanguo jingji huiyi zhuankan* [Special Collection of Proposals from the National Economic Conference] (n.p, 1928), passim.
8. Jiang Jieshi was directly involved in appointing provincial salt commissioners, forbidding soldiers to smuggle salt, and deciding how to raise revenue from the 20 million *dan* of salt inventory held at Haizhou worth 60 million *yuan*. See entries dated February 23 and October 17, 1928 in Wang Zhenghua, ed. (2011-).
9. Jing's letter dated May 28, 1927 and Sun Zhongxian's letter to Fan Xudong dated June 9, 1927 in *JD* 700259–260 and 700262 respectively.
10. Shanghaishi dang'an'guan, ed., *1927 nian de Shanghai shangye lianhehui* [The Shanghai Chamber of Commerce, 1927] (Shanghai: Renmin chubanshe, 1983), 117, 125; Ding Changqing (1988), 41–48; and Jiang Liangqin, *Nanjing guomin zhengfu neizhai wenti yanjiu* [Nanjing Government's Problem of Domestic Debt] (Nanjing: Nanjing daxue chubanshe, 2003), 48, 143.

11. Jiang's confidential cable to Fu Zuoyi dated October 29, 1928 and pleas on behalf of the Changlu revenue farmers in *CLHB* 680.9.502.
12. Adshead's assessment of the inspectorate's enduring strength might thus be optimistic. See Adshead (1970), 197. Eventually, a revamped Salt Administration incorporated the inspectorate, with foreign staff retained on condition that they served only the Finance Ministry, not the bank group or foreign bondholders. See Song Ziwen's confidential letter to chief inspector dated August 15, 1932 in Arthur N. Young Papers, Box 32.
13. Xu Zhaozhou to Fan Xudong dated August 18, 1928 in *JD* 700303.
14. Letter of Fu Bingzhi (who had studied at Harvard) to Fan Xudong dated August 18, 1930.
15. Indeed, it became part of the gabelle even after reunification of the country. See Finance Ministry order issued July 30, 1928 in *ZGJDYWSZL* II: 189. Zhou, Jing's former colleague, was then Zhejiang provincial salt commissioner. He had served with Zou in Canton. See Jing Benbai letter to Fan Xudong dated July 5, 1928 in *JD* 101362.
16. Caizhengbu Yanwushu, ed., *Yanwu nianjian* [Annual Report on Salt Affairs] (Nanjing: Caizhengbu Yanwushu, 1930), 113; and Caizhengbu Yanwushu, comp., *Yanwu taolunhui huiyi huibian* [Proceedings of the Conference on Salt Affairs] (Nanjing: Ministry of Finance, 1928), passim.; *JD* 101369, and Jing (1935), 22.
17. Contract signed December 15, 1928. Ding Changqing (1988), 41.
18. Caizhengbu Yanwuchu, comp. (1930), *yunxiao*, 15–44; and Zhang Lijie (2011), 157–63.
19. Ding Changqing and Tan Renyue, eds., *Zhongguo yanyeshi* [A History of China's Salt Industry], 3 vols. (Beijing: Renmin chubanshe, 1997), III: 142–3. After three years in office, Zou allegedly had amassed a personal fortune, with 300,000 *yuan* from the reregistration alone. See Jing's letter to Fan Xudong dated 1931 in *JD* 700373.
20. *ZGJDYWSZL* II: 171–2. See also Liu Jun (1934), 116–35; Ding Changqing, et al. (1990), 218–44; and Strauss (1998), 89.
21. Minsheng, registered in 1927, spent three years securing authorization to ship its product in gunny bags. See its petition to the Liangzhe commissioner dated June 15, 1929 in Papers of the Liangzhe Salt Administration L057-14-91 held at Zhejiang Provincial Archives, Hangzhou.
22. *JD* 100174–76.
23. Ding Changqing and Tang Renyue (1997), 145. For the regulations, see *ZGYZSL* IV: 90–101.
24. Hongyuan et. al. petition to the Finance Ministry dated November 30, 1931 in *JD* 100198–100204.
25. NSRB petition to the Finance Ministry dated February 4, 1931, and Song Ziwen's reply dated July 15, 1931 in *JD* 100187–100193.
26. Chen Changheng, "Xin yanfa de qicao jingguo ji qi neirong shuoming," [The Drafting Process of the New Salt Law and an Explanation] *YZZZ* 52 (1931), *xuanlun* [Selected Discussion], 1–17; and Jing Benbai (1935), *zalu*, 13–16. Jing served as Zhuang's secretary at the Zhejiang provincial salt administration in 1911. See Jing Benbai ed., *Yanwu gemingshi* [A History of

the Revolution in Salt Administration] (Beijing: Yanzheng zazhishe, 1929), 7. Liu was a shareholder of both Jiuda (36 shares) and Yongli (40 shares). Chen, Wei, and Sun Xuewu were classmates at the Qinghua Preparatory school (under the supervision of Fan Yuanlian), University of Michigan and Harvard. See Wei Tingsheng, *Wei Tingsheng zhizhuan* [Autobiography of Wei Tingsheng] (Taibei: Zhongwai tushu chubanshe, 1977), 32, 102.
27. Jing Benbai's letter to Fan Xudong dated April 2, 1931 in *JD* 101401.
28. Chen Changheng (1931), 7; and Benbai, "Xin yanfa shixing zhi anjiao" [Hidden Shoals in Implementing the New Salt Law] *YZZZ* 52 (1931), editorial, 1–5.
29. "The Problem of Putting the Principle of the New Salt Law into Effect," F. A. Cleveland memo to the Finance Ministry, Part I, in Arthur N. Young Papers, Box 116.
30. Letters to L. C. Wolcott dated May 6, 1927 and March 31, 1929 in R. D. Wolcott Letters.
31. For arguments from both sides, see *DGB* March 15, 1931; *DGB* March 27, 1931 interview with Jing Benbai; rebuttals from Shandong revenue farmers on April 6, 1931; April 15, 1931 from Changlu; further comments on the high price of salt on April 23, 26–27, and 29, 1931; and June 1, 1931 and July 7, 1931 against the inspectorate. See also Ding Changqing, ed. (1990), 175–8.
32. According to Cleveland, even at that rate revenue would be reduced by 14,558,000 *yuan*, leaving approximately 4.2 million *yuan* to the central government each month. See his memo "On the estimated returns from duty rate of $8.00 per 100 kilogram and comparison with estimated return for the 20th fiscal year" in Arthur N. Young Papers, Box 32.
33. Liu Cunliang, "Zhongguo renmin zhi yanshui fudan," [The Salt Tax Burden Borne by the Chinese People] *Zhongguo jingji* [China's Economy] 2.2 (1934), 1–40, esp. 38.
34. Ma Yinchu, *Caizhengxue yu Zhongguo caizheng* [Fiscal Science and China's Public Finance], 2 vols. (Shanghai: Shangwu yinshuguan, 1948), I, 199.
35. Neither side got all it wanted when the commission was established on January 22, 1935. Caizhengbu caizheng kexue yanjiusuo and Zhongguo di'er lishi dang'anguan, comp., *Guomin zhengfu caizheng jinrong shuishou dang'an shiliao, 1927–1937* [Archival Materials on Fiscal, Financial and Taxation History of the Nationalist Government, 1927-1937] (Beijing: Zhongguo caizheng jingji chubanshe, 1997), 836–7, 839–43; and Zhang Lijie (2007), 15. On lobbying to pack the commission by both sides, see Zhong's letter to Fan Xudong dated June 19, 1934 in *JD* 701534, 701541–2 and *DGB* May 11, 1935.
36. The Lianghuai merchants alone borrowed from various banks over 34 million *yuan*. See Shanghai Commercial Savings Bank, ed., *Yanye fangkuan huibian* [Consolidated Report on Salt Loans, 1934], Q275.1.1632 held at *SM*. With substantial exposure to the revenue farmers in their loan portfolios, Zhang Gongquan (general manager, Bank of China) and Hu Bijiang's (general manager, Bank of Communications) appointment to the

commission thus posed a potential conflict of interest. See Zhong Lijian's letter to Fan Xudong dated January 24, 1935 in *JD* 701543. For a balanced discussion of the issues, see Zhao Shuyong, "Zhongguo yanwu zhi nianjian," [An Overview of China's Salt Affairs] in Jiang Fenglan, *Guomin zhengfu shiqi de yanzheng shiliao* [Historical Materials on Salt Administration under the Nationalist Government] (Taibei: Academia Historica, 1993), 363–384, esp. 380–3.

37. 1933 cost per *dan* for Changlu, 0.26 yuan; Shandong 0.31; Lianghuai, 0.67, and Northern Sichuan, 6.84. See Zhang Wei, "Zhongguo zhanshi shiyan tongzhi wenti," [Wartime Problem of Controlling Alimental Salt] *Dongfang zazhi* [Eastern miscellany] (Aug., 1936), table 1.
38. Chen Lifu, "Wo de chuangzao changjian yu fuwu," [My Innovations, Proposals, and Services] *Zhuangji wenxue* [Biographical Literature] 55.1 (1989), 108. Chen was a member of the commission and opposed the law on strategic grounds, arguing that Changlu, with the lowest production cost by far, would be the only viable production area left under free trade.
39. *Jingyan dengji'an shimo* [Registration of Salt Refineries: A Documentary History] (N.p., 1934), 1–7. Cai Guoqi, the Changlu Revenue Inspector, opposed the extension of reregistration to the refineries as legally indefensible. See Jing Benbai's letter to Fan Xudong dated January 28, 1933 in *JD* 101536 and 101539.
40. Jing Benbai's letter to Fan Xudong dated January 2, 1933 in *JD* 101530–1.
41. Zhong Lijian's letters to Fan Xudong dated January 19, 1933 and January 28, 1933 in *JD* 101532 and 101542. On their service records at the Salt Administration, see S-2–138 held at *AS*. Zhong (1876–1966) served at the Revenue Inspectorate from 1921 until 1924 when he began working for Jing and Fan.
42. Jiuda petition dated February 6, 1933 in *JYDJ*, 15–17.
43. Fan Xudong's cable to Song Ziwen in *JD* 101540–1.
44. Thus avoiding "unnecessary competition" among the refineries, a point made by Jing Benbai. See his letter to Fan dated January 28, 1933 in *JD* 101536–8.
45. Zhong Lijian's cable to Fan Xudong dated February 3, 1933 in *JD* 101546. On Zhu, see his Salt Administration Service Record 20733, S-02–103 held at *AS*.
46. Fan Xudong's cable to Jiuda head office dated February 12, 1933 in *JD* 101466.
47. The dispute centered over whether the advance could be paid in promissory notes (as had been the standard procedure) to the inspectorate, or in cash to the commissioner's office. Jing Youyan insisted on cash and in one lump sum to his office, while Cai sided with the company (and hence into Nanjing's account). See Jing's order dated March 2, 1933 and threat to shut down the company dated March 7, 1933 in *JD* 101468; Cai's advice dated March 17, 1933 in *JD* 101617; and Fan's petition hand-delivered to Zeng Yangfeng requesting his intervention with Jing dated March 14, 1933 in *JD* 101472.
48. Fan Xudong's company circular dated May 8, 1933 and Jing Benbai's letters to Fan Xudong dated May 23, 1933 in *JD* 100518–20 and 101708.

49. Zhou's cable dated April 15, 1933 and Song referring the matter on to his subordinates two days later in Jincheng Bank papers, Q264.1.670 held at *SM*.
50. Letter dated April 3, 1933 in *JD* 101628.
51. Zhong Lijian letter to Lin Ziyou dated May 6, 1933; and cable from Fan to Jing Benbai in *JD* 101664, 101680.
52. On Dinghe and its board chairman, Zhong Fuqing, see the company's record held at *AS*.
53. Transcript of NSRB meeting in *JD* 101719.
54. Zhong Lijian's letters to Fan Xudong dated June 2 and 6, 1933 in *JD* 101733, 101722–25.
55. Report # 2875 on Wuhe Refined Salt Co., Ltd., dated September 30, 1933, Bankers' Cooperative Credit Service, Q275.1.1991 in *SM*.
56. Contract dated June 15, 1933 with Jincheng, Zhongnan, and National Commercial Bank in *JD* 101508–9.
57. On the negotiations and Fan Xudong's counteroffer of 0.7 yuan per *dan* to Zhu Tingqi, see *JD* 101484. Revenue farmers paid fees ranging from 0.42 *yuan* (Liangzhe) to 0.68 *yuan* (Lianghuai), and 0.87 *yuan* (Jiangxi) per *dan*. See Zhong Lijian's cable to Fan Xudong dated May 7, 1933 in *JD* 101673.
58. Letters from Zhong and Chen Canglai to Fan Xudong in *JD* 701385 and 700742, respectively. Dinghe went bankrupt before paying the reregistration fee. See *ZGYZSL*, 2nd comp. (1943), I: General introduction, 20.
59. Letters from Zhong Lijian to Fan Xudong dated July 16–25, 1933 in *JD* 701387–89.
60. Zhong Lijian's letter to Fan Xudong reporting on his meeting with Xia Zhubo, representative of the Yingkou refineries dated September 16, 1933 in *JD* 701404.
61. Order issued October 16, 1933 in *JD* 700620.
62. Kōgyō kagaku kai Manshū shibu, ed., *Manshū no shigen to kagaku kōgyō* [Resources and Chemical Industries in Manchuria] (Tokyo: Maruzen kabushiki kaisha, 1937), 286; and Kunio Suzuki, *Manshū kigyōshi kenkyū* [A Business History of Manchuko] (Tokyo: Nihon keizai hyōronsha, 2007), 612.
63. Lin Shougu, "Jingyanye zhi weiji ji qi wanjiu banfa" [The Crisis Confronting the Refined Salt Industry and Solutions] *Haiwang* 7.22 (1935), 388. On the smuggling of Yingcheng gypsum salt, see *ZGYZSL* (1933), I: E'an, section 6, 26; and Chen Jiaren, "Yingcheng gaoyan chanxiao jianshi," [A Brief History of the Production and Marketing of Gypsum Salt from Yingcheng] *Hubei wenshi ziliao* [Historical materials of Hubei] 10 (1984), 158–73.
64. Zhong Lijian's letters to Fan dated January 4 and 16, 1934 in *JD* 701414 and 701419.
65. Fan was nominated to the committee by Qian Changzhao (1899–1988). See Qian's memoir, *Qian Changzhao huiyilu* (Beijing: Zhongguo wenshi ziliao chubanshe, 1998), 37–8.
66. Zhong's letters to Fan Xudong dated March 30, 1934 in *JD* 701442–3, 701448.
67. Zhong's report to Fan dated June 13, 1934 in *JD* 701465–6.
68. For a partial listing of such seizures and persecutions after June 1933, see *Jingyan dengji"an shimo* (1934), 84–87.

69. For He Jian's proclamation, see *Haiwang* 7.9 (1934), 151.
70. Copy of Fan Xudong's cable to Kong Xiangxi dated November 23, 1934 in *JD* 200786. A secretary at the ministry leaked Kong's order reprimanding He, giving the industry time to draft for He a preempting cable seeking authorization from Jiang Jieshi. See *JD* 701494–701058.
71. Fan's petition dated September 16, 1934 in *JD* 200747. See, for examples, Zhenjiang October 24 and 28, 1934; Huaining, October 25, 1924; Wuxi, November 5, 1934; Jiujiang, November 9, 1934 and *Haiwang* 7.9 (1934), 150–1.
72. As a last resort, Zhong planned a march on Nanjing of thousands of workers from the refineries, supplemented by consumer groups in support of the "New Life Movement." See *JD* 200761 and 701843–7.
73. Sent 29 October 1934 and November 13, 1934 in *JD* 200768.
74. Zhong's letter to Fan dated November 16, 1934 in *JD* 701504.
75. See Chen's service record at the Salt Administration, and Yongli shareholders list (1917–1922), both held at *AS*. He was a member of the 1910 class of Boxer Indemnity scholars, counting among his classmates Hu Shi and Zhao Yuanren.
76. Text of Zhong's presentation in *JD* 100256–62.
77. Zhong might be exaggerating the industry's imminent bankruptcy, but the fee would certainly took the lion's share of Jiuda's 0.602 yuan net profit per shipped *dan* if the company could not raise its price.
78. Zhong Lijian's letter and cable to Fan Xudong dated November 27, 1934 in *JD* 701515 and 701519.
79. Fan Xudong's cable to Zhu Tingqi, November 28, 1934 in *JD* 701521.
80. Chen followed Ma Taijun to Shanghai to ensure that Zhu would not change his mind at the last minute. If he did, Chen vowed to resign. See Zhong's letter to Fan Xudong dated December 5, 1934 in *JD* 701528.
81. *ZGJDYWSZL* II, 184–6.
82. Yihexiang, the monopoly holder of Nanjing, went bankrupt after several deaths were traced to contaminated crude salt in May 1934.
83. Originally Chen Guangfu, general manager of the bank and a leading banker, was to make the call. See *JD* 701532. Zou had worked closely with Fan Xudong on the Nitrogen Fertilizer Project under Kong's watch at the Ministry of Industries.
84. For a list of subsidized newspapers, see *JD* 701504. Eventually, the industry operated its own Continental News Agency (Dalu xinwenshe).
85. See, respectively, *JD* 701472 and 701545.
86. Guojia shuiwu zhongju, comp. (1999), 136–40. Taking into account the conversion in 1934 from *sima* scale (127 *jin* per *dan*) to metric (*shi*) scale (100 *jin* per *dan*), the average gabelle rate by weight was increased by another 22 percent (Cleveland's estimate) or 27 percent (Jiuda's estimate), to 6.8 *yuan* per *dan*.
87. Adapted from Tables 2.5/6 in Dong Zhengping, *Kangzhan shiqi Guomin zhengfu yanwu zhengce yanjiu* [Salt Policy of the Nationalist Government during the War of Resistance] (Ji'nan: Qilu shushe, 2004), 75–6.

88. A negative −0.25 correlation coefficient between revenue collected and shipped salt suggests that the rising revenue had more to do with rate increases than actual shipment.
89. For the array of subsidies paid by both sides uncovered in 1931–2, see Guojia shuiwu zongju, comp. (1999), 149–53.
90. To the extent that Zhu consulted Zhong Lijian on how to lobby the Legislative Yuan and even had the NSRB retain the Royal Detective Agency (Huangjia zhentanshe) in Shanghai to monitor the activities of his rivals in 1936. See *JD* 701621 and 701635.

7 At War

1. Fan Xudong letter to Hu Shi dated October 20, 1938, in Hu Shi Archives held at the Special Collections, Institute of Modern History, Chinese Academy of Social Sciences, Beijing.
2. Odoric Y. K. Wou, "The Phenomenon of Collaboration," in David Pong, ed., *Resisting Japan* (Norwalk, CT: Eastbridge, 2008), 175–205, esp. 190; and Brett Sheehan, "When Urban Met Rural," paper presented at the conference on "Daily Life in Wartime China", U.C. Berkeley, 2009.
3. Parks Coble, *Chinese Capitalists in Japan's New Order* (Berkeley, CA: University of California Press, 2003), 168–75; Weng Wenhao, "Kangzhan ernianlai de jingji," in Shishi wenti yanjiuhui, comp. *Kangzhan zhong de Zhongguo jingji* [China's Economy in the War of Resistance] (n.p.: Kangzhan shudian, 1940), 127; Shen Zuwei and Du Xuncheng, eds., *Guonanzhong de Zhongguo qiyejia* [Entrepreneurs during China's National Crisis] (Shanghai: Shanghai shehui kexueyuan chubanshe, 1996), passim., and Dai Angang, "Kangzhan shiqi Zhongguo xibu gongye de zhuanji he kundun," [Opportunities and Limitations of Industrial Development in Western China during the War of Resistance] in Zhu Yingui and Dai Anguan, eds., *Jindai Zhongguo* [Modern China] (Shanghai: Fudan daxue chubanshe, 2006), 672–93.
4. Kishi Toshihiko, "Yongli kagaku kōgyō konsu to Fan Xudong," in Soda Saburō ed., *Chūgoku kindaika katei no shidōsha tachi* [Leaders in China's Modernization Process] (Tokyo: Tōhō shoten, 1997), 254–85. On the porous boundary between collaboration and resistance, see Timothy Brook, *Collaboration: Japanese Agents and Local Elites in Wartime China* (Cambridge, MA: Harvard University Press, 2005); and Zanasi (2006).
5. The lineage of work unit (*danwei*) as a feature of state enterprise in modern China can also be traced to private enterprises such as Jiuda during the republican period. See Morris Bian, *The Making of the State Enterprise System in Modern China* (Cambridge, MA: Harvard University Press, 2005) and Joshua H. Howard, *Workers at War* (Stanford, CA: Stanford University Press, 2004), 363.
6. Li Zhuchen's undated petition to the Executive Yuan, ca. July, 1946, in the archives of the Jieshou gongzuo weiyuanhui, Ministry of Economics, 18–36a-3,

held at *AH*. The contract with Otani was backdated to January 5, 1937. See *JD* 200641–2.
7. Petition made August 1, 1937. See Fan' summary of the events dated November 15, 1938 in Ministry of Economics 18–06–34 held at *AS*.
8. Qian Changzhao, "Guomindang zhengfu ziyuan weiyuanhui shimo," [On the Nationalist National Resources Commission] in Quanguo zhengxie wenshi ziliao yanjiu weiyuanhui gongshang jingjizu, ed., *Huiyi guomindang zhengfu ziyuan weiyuanhui* [Remembering the Nationalist Government's National Resources Commission] (Beijing: Wenshi chubanshe, 1998), 160. For the relocation of Shanghai industries, see Hu Juewen, *Hu Juemin huiyilu* [Memoirs of Hu Juemin] (Beijing: Zhongguo wenshi chubanshe, 1994), 50; and Huang Liren, *Kangzhan shiqi dahoufang jingjishi yanjiu* [An Economic History of the Interior During the War of Resistance] (Beijing: Zhongguo dang'an chubanshe, 1998), 150–1.
9. The evolution of Sichuan's salt administration from nominal "free marketing" to designated merchant and back is too complex a story to tell here. See Lin Zhenhan, *Chuanyan gaiyao* [A Concise History of Sichuan Salt], rev. ed. (n.p.: Shangwu yinshuguan, 1919), 33–62; ZGYZSL II, "Chuannan," 1–128, "Chuanbei," 1–112; Miao Qiujie et. al., *Chuancuo gailue* [An Overview of Sichuan Salt] (Np., 1939), passim.; Li Han, *Miao Qiujie yu Minguo yanwu* [Miao Qiujie and Salt during the Republican Period] (Beijing: Zhongguo kexue jishu chubanshe, 1990), 104–11; Zigongshi Yanwu guanliju comp., *Zigongshi yanyezhi* [A History of Zigong Salt Industry] (Chengdu: Sichuan renmin chubanshe, 1995), 234–36; Zhongguo yanye zhonggongsi, ed. (1997), 148; and Madeleine Zelin (2005).
10. Caizhengbu Yanwu zongju, comp., *Yanwu faling huibian* [Compendium of Salt Laws and Regulations] (Chongqing: Caizhengbu Yanwu zongju, 1942), 14–16, 24–26, 35–40; and Dong Zhenping, *Kangzhan shiqi guomin zhengfu yanwu zhengce yanjiu* [Nationalist Government's Salt Policy during the War of Resistance] (Ji'nan: Qilu shushe, 2004), 164. By 1941, gabelle revenue collected had fallen to 100 million yuan from the prewar high of 240 million yuan. Letter from Wolcott to Cox dated June 19, 1941 in Arthur N. Young Papers, Box 74.
11. "Wo'men chudao Huaxi," [Our Arrival in Western China] repr. in *Huagong xiandao Fan Xudong* biansheju comp., *Huagong xiandao Fan Xudong* [Fan Xudong: Pioneering Industrial Chemist] (Beijing: Zhongguo wenshi chubanshe, 1987), 22–3; and "Chunyu linlizhong de yige shenghui," [A Magnificent Event in Spring Showers] *Haiwang* 16.24 (1944), 186. Less than hospitable reception of other evacuees was analyzed in Sun Guoda, *Minzu gongye daqianxi* [Mass Migration of the National Bourgeoisie] (Beijing: Zhongguo wenshi chubanshe, 1991), 30–36.
12. Fan's proposal dated March 14, 1938. See *JD* 300488–9.
13. Fan's cable to Miao Qiujie dated March 21, 1938 in *JD* 300487.
14. *JD* 300491–2.
15. Recommendation of provincial salt administration and approval by the Salt Administration dated April 12, 1938, *JD* 300497–8.

16. Fan's letter to Lu Zuofu dated March 7, 1938; Lu's letter to Xu Kan two days later; Weng Wenhao's letter to Xu Kan dated March 25, 1938 in 256.1866-1 held at*AH*; and Fan's appeal to Jiang Jieshi dated March 23, 1938 and to Kong Xiangxi on April 5, 1938 in *YL* 300985 and 300990 respectively. Fan, Lu and Zhou Zuomin all served on the board of the Zhongghua Shipbuilding and Engineering Co., Ltd. (Shanghai).
17. On the Dapu plant launched in late 1936 at the suggestion of Zeng Yangfeng, see Tang Hansan, "Huaichang huiyilu," [Memories of the Huai Factory] *Haiwang* 16.31 (1944), 249–51.
18. Xue Xianzhi (1992), 73.
19. Petition by salt merchants dated April 23, 1938, *JD* 300495-6.
20. Confidential cable to Chongqing Military Commission from Wang et al., dated May 11, 1938, *JD* 300501-2. The love–hate relationship between the salt merchants and Sichuan's warlords is analyzed in Song Liangxi, "Sichuan junfa dui Zigong yanshang de jielue," [Sichuan Warlords and Zigong Salt Merchants] *Jingyanshi tunxun* [Bulletin on the History of Well Salt] 1 (1982), 21–7.
21. Tang Hansan, "Zaijing liunian," [Six Years at Zigong] *Haiwang* 16.31 (1944), 245.
22. Xue Xianzhi (1992), 72.
23. Cable dated May 11, 1938 in *JD* 300501.
24. Xue Xianzhi, "Jiuda Yongli qianchuan yihou," [Jiuda and Yongli in Sichuan] in Tan Jie et al., eds., *Hongsanjiao de huihuang* [Glory of the Red Triangle] (Tianjin: Xinhua tongxunshe, 1997), 115.
25. Zhong Lijian's cable to Li Zhuchen dated June 30, 1938, *JD* 300515.
26. Xie had shared the cable from Wang et al. with the company. See *JD* 300514, 300544-6, and Xue Xianzhi (1997), 115.
27. Speech given on opening day, September 18, 1938, in *JD* 300523-4.
28. Tang Hansan (1944), 245.
29. On the history of this cartel, see "Furong yanchang zengchan yijianshu," [Recommendations on Increasing Production at the Furong Saltyard] in *JD* 300755-6.
30. Jiuda petition to the Sichuan Salt Administration dated August 21, 1938, *JD* 300516.
31. Tang Hansan (1944), 246; and Wu Peiji, "Zigong yanhua gongye fazhan jianshi," [A Brief History of the Development of Salt-based Chemical Industry at Zigong] *Zigong wenshi ziliao* [Historical Materials of Zigong] 16 (1986), 81.
32. Tang Hansan cable to Kong Xiangxi and Weng Wenhao dated September 18, 1938 in *JD* 300522; Ministry of Finance order dated October 11, 1938, *JD* 300535.
33. Finance Ministry order dated October 24, 1938, *JD* 300537. On the myriad of surcharges and levies, see Zhong Changyong, "Si'nan junfa yu Sichuan yansui," [Warlords in the Southwest and Sichuan's Gabelle] *Jingyanshi tunxun* 1 (1984), 39–46.
34. Petition dated October 20, 1938, *JD* 300532-4.
35. Plan dated April 10, 1940, *JD* 300554-55.

36. Daily log of Jiuda West China office, dated February 4, 1939 and March 13, 1939, JD 301143 and 301152, respectively. See also "Papers of Dingchang Salt Company," JD 100755–6.
37. "Sichuan Zigong Ziliujingchang chuangban jingge," dated 1941, JD 300565.
38. Daily log of Jiuda West China office, dated February 5, 1939, JD 301143. Miao was removed from office later that year. See Li Han (1990), 124–5.
39. "Report to shareholders, 1941," in JD 300566.
40. JD 300628–9.
41. Summary on the Hunan market, 1938–1943, Minutes of managers' meeting dated October 7, 1944, JD 300631.
42. Sun Xuewu's letter to Lu Zuofu dated January 2, 1941; Lu's letter to his brother dated January 15, 1941; and a favorable reply ten days later. See JD 800351, 800349.
43. Zelin (2005), chap. 8.
44. JD 300597, 300628, 300615, 300645.
45. Tang Hansan letter to Weng Wenhao dated September 19, 1938 in JD.
46. Fan Xudong's speech dated July 20, 1939 commemorating the 25th anniversary of Jiuda. Reprinted *Zigong wenshi ziliao* 15 (1985), 126.
47. Lu Bo and Liu Jiashu, "Zhitiaojia zhi xingneng ji yanlu nongsuo shiyan," [An Experiment to Increase Brine Salinity Using Bamboo Shed] *Huanghai huagong huibao* 1.1 (1940), 16– 29; Tang Hansan, "Zaijing sinian" [Four Years at Zigong], JD 300586–91.
48. JD 300588, 300591; Zhao Zhi, "Zigong yanchang dianli tuilu jiche de changshi," [Electrification of Brine Pumping at Zigong Saltyard] *Zigong wenshi ziliao* 15 (1985), 170–5.Large Zigong brine wells had begun experimenting with steam-engine powered pumps since the early twentieth century. See Zhong Changyong, "Furong yanchang de jiche jilu," [Mechanized Brine Pumping at Furong Saltyard] in Zigongshi yanye lishi bowuguan, ed., *Sichuan jingyanshi Luncong* [Essays on the History of Sichuan Well Salt] (Chengdu: Sichuansheng shehui kexueyuan chubanshe, 1985), 332–44.
49. Liu Jiashu and Lu Bo, "Gaijin Jinle huayanzhao zhi" [An Improved Furnace Design for Jinle Saltyards], *Huanghai huagong huibao* [Bulletin of the Golden Sea Chemical Industry Research Institute] 1.1 (1940), 3–6.
50. Zhong Zihuang, Liu Jiashu, and Lu Bo, "Muzhazhi yanzhuan zhi jingguo," [Compressing Salt Brick with Wood Mold] *Huanghai huagong huibao* 1.1 (1940), 7–15; and order to Jiuda dated May 21, 1941, in "Ministry of Finance Archives" 266.339 held at the Second Historical Archives, Nanjing.
51. Huanghai huaxue gongye yanjiushe, ed., *Huanghai huaxue gongye yanjiushe ershi zhounian jininace* [Twentieth Year Anniversary Commemorative Volume of the Golden Sea Chemical and Industrial Research Institute] (n.p., 1942), 30.
52. JD 300575–6; Zhao Zhi, "Zigong yanchang dianli tuilu jiche de chuangshi," [Pumping Brine using Electricity at Zigong Salt Works], *Zigong wenshi ziliao* 15 (1985), 171. Both Esson Gale and Arthur Young had visited the plant.

53. Zhao Boquan, "Luhuabei yu shiyan zhi fenli," [Separation of Barium Chloride from Alimental Salt], *Huanghai yanjiu diaocha baogao* [Survey Report of the Golden Sea Chemical Industry Research Institute] 32 (1942).
54. Sun Jishang, "Jianle yanchang ganzhi fenxi," *Huanghai huagong huibao* 1.1 (1940), 30–4; and Wu Peiji, "Zigong yanhuagong fazhan jianshi," [A Brief History of the Development of Zigong Chemical Industry] *Zigong wenshi ziliao* 16 (1986), 83.
55. For Sanyi's charter and registration on September 6, 1943, see company registers, Ministry of Industries Archives 18–23–0-74–37–000–002 held at *AS*.
56. Wang Wenda, "Sanyichang yu Gongjing yanchan zhi qiantu," [Sanyi and the Future of Zigong] *Haiwang* 30 (1943), 206–7. The company had hoped to realize a 30 percent profit. See *JD* 300624.
57. For a list of these chemical works, see Wu Peiji (1986), 93–4.
58. Xu Dingxin, "Fan Xudong," in Shen Zuwei and Du Xuncheng, eds. (1996), 145.
59. Rescripts of the Changlu Salt Revenue Administration held at the Institute of Economics, Nankai University.
60. Estimated production cost per *dan* at 3.315 yuan, October 7, 1938, *JD* 300614.
61. *JD* 200001.
62. *JD* 300600.
63. Song Shangze and Wang Zhenhua, "Kangzhan shiqi chuanyan de jiage guanli," [Price Control on Sichuan Salt during the War of Resistance] *Zigong wenshi ziliao* 16 (1986), 67.
64. *JD* 300570.
65. *JD* 300567, 300645.
66. Jiuda and Yongyu completed its shipment of "ever normal" salt reserve in November 1936, which cost the companies 1.7 million yuan in gabelle advance, again underwritten by Jincheng. See *JD* 200221 and Q264.1.690 held at *SM*.
67. Dong Zhenping (2004), 129–40. Beginning in 1939, sale of tobacco and alcohol came under state control. In addition to salt, sugar, matches, antimony, tungsten, and tea soon followed. See He Shimi, *Kangzhan shiqi zhuanmai shiliao* [Historical Materials of State Monopolies during the War of Resistance] (Taibei: Academia Historica, 1992), passim.
68. Miao Qiujie and other officials first proposed state control of salt production and marketing in late 1935. See summary of meeting leaked to Jiuda in *JD* 700689, 701528–30.
69. "Memo on nationalization from the Salt Administration to the Ministry of Finance" dated July 28, 1937 in *ZGJDYWSZL* IV: 10–11, 29–31; Salt Administration confidential memo to the Finance Ministry in December 1940 in Arthur N. Young Papers, Box 74; and Guojia shuiwu zhongju, comp.(1999), 188–92.
70. *ZGJDYWSZL* IV: 38–42.
71. "Memo on the proposed salt monopoly in China" by O. C. Lockhart dated February 27, 1941 in Arthur N. Young Papers, Box 74. Lockhart also noted that 65 percent of the bonds were actually owned by Chinese interests.

72. Dated May 13, 1938 in R. D. Wolcott letters.
73. Wolcott's confidential letter dated June 19, 1941 in Arthur N. Young papers, Box 74. Wolcott, too, resigned in February 1942 after Kong Xiangxi denied him a raise.
74. On Japanese use of occupied China's economy for its own purposes, see Kōain kahoku renrakubu, *Kahoku engyō rutchi jōken chōsa hōkokusho* [Survey Report of Production Condition of Salt in North China] (n.p., 1941), passim.; Shingo Sumiyoshi, et al., *Chūka engyō jijō* [Salt in China] (Kawasaki: Ryūshuku sanbō, 1943), 438–9; Kōain kachū renrakubu reizai dai 2-kyoku, *Kachū engyō kohin yūgen kōshi ni tsuite* [Central China Salt Industries, Ltd.] (Nanjing: n.p., 1941), 3–4; Coble (2003); and Stephen R. Mackinnon, Diana Lary, and Erza Vogel, eds., *China at War* (Stanford, CA: Stanford University Press, 2007), passim.
75. Although not without debate among academic economists such as Ma Yinchu, He Lian, and Fang Xianting. See Huang Lingjun, "30–40 niandai Zhongguo sixiangjie de jihua jingji sichao," [Planned Economy as a Concept among Chinese Intellectuals during the 1930s and 40s] *JDSYJ* 2 (2002), 150–76. On state control of ordnance and other heavy industries in this period, see Howard (2004) and Bian (2005).
76. Ding Changqing, et al. (1990), 350; and Dong Zhenping (2004), 214.
77. Minutes of managers meeting, November 6, 1943, *JD* 300621, 300632.
78. *JD* 300616.
79. See Yanwu zongju, comp. (1942), 55; Caizhengbu Yanwu zongju (1944), 55, 120–1, 130, 140; and Yang Xingqin, *Zhongguo zhanshi yanwu* [Salt Affairs in Wartime China] (n.p.: Guomin chubanshe, 1943), 72–6.
80. *JD* 300572, 300633.
81. Rulings dated April 21, 1943 and April 6, 1944 in Ministry of Finance Archives 266.168 held at the Second Historical Archives, Nanjing.
82. Weng Wenhao, *Kexue yu xiandaihua* [Science and Modernization] (Beijing: Zhonghua shuju, 2009), 544; *YL* 300493, Fan's letter to Yongli's Chongqing office dated April 18, 1938; and Ministry of Finance letter to the company dated April 23, 1938 in *YL* 300965 and 300991.
83. Order to file a new company charter was issued by Chen Lifu and Xu Kan. See Ministry of Economics 18–06–34 held at *AS*.
84. *AS* 256.1866-2, Fan's cable to Kong dated May 3, 1938. Comments by Xu Kan and Zou Lin, vice ministers, on memo of the Treasury Bureau dated May 9, 1938, noting that Weng had phoned to request a delay until he had discussed the matter with Fan. See *AS* 18–72–257–1.
85. Fan's cable to Lu Zuofu and He Lian dated 7 June 1938; Fan's cable to Xu Kan and Kong dated June 9, 1938; and memo from the Treasury Bureau to Kong dated June 13, 1938 in *AH* 256.1866-2. Final installment received on March 13, 1939, see *YL* 301004/301133.
86. Shen Zuwei and Du Xuncheng eds., (1996), 38–41.
87. Generalissimo's Office personnel file 12957 in *AH*; and William C. Kirby, "China's Wartime Economy," in James C. Hsiung and Steven I. Levine, eds., *China's Bitter Victory* (New York: M. E. Sharpe, 1992), 185–212.

88. On the Universal Trading Corp., created as a front to help China channel credit from the U.S. Export-Import Bank, see Box 4, K.P. Chen papers held at Special Collections, Columbia University.
89. Archives of the National Government (headquarters) 1120.1032.1 held at *AH*.
90. Chen Bulei's comments dated October 1, 1943, and Jiang's decision dated October 7, 1943 in ibid.
91. Jiang Jieshi's memo to Kong Xiangxi and Weng Wenhao dated October 7, 1943 in Ministry of Economics Archives 18–22–259–1 held at *AS*.
92. Qian Changzhao's memo to Jiang dated October 15, 1943 in Archives of the National Government (headquarters) 1120.1032.1 held at *AH*.
93. Ling Yaolun and Xiong Pu, eds., *Lu Zuofu wenji* [Collected Works of Lu Zuofu] (Beijing: Beijing daxue chubanshe, 1999), 570. The delegates lunched with Jiang on September 19, 1944. Song Ziwen, Weng Wenhao, and Jiang Weiguo joined them. See diary entry, Box 8, file 1, in K. P. Chen Papers.
94. Diary entry October 21, 1944, Box 7, file 2, K.P. Chen Papers. Chen was reassuring a concerned Lauchlin Currie, Franklin Roosevelt's economic adviser.
95. *Ibid*. See also George C.X. Wei, *Sino-American Economic Relations, 1944–1949* (Westport, CT: Greenwood Press, 1997), 8.
96. Olin S. Pugh, *The Export-Import Bank of Washington* (Columbia, SC: University of South Carolina Bureau of Business and Economic Research, 1957), 8.
97. Archives of the National Government (headquarters) 1120.1032.1 held at*AH*.
98. Weng's recommendation dated February 16, 1945; Chen's endorsement two days later in the Archives of the National Government (headquarters) 1120.1032.1 held at *AH*; and Jiang's approval dated March 4, 1945 in Ministry of Economics Archives 18–22–259 held at *AS*.
99. O. C. Lockhart to Emilio (Pete) Collado dated 25 January 1945; John Carter Vincent to Lockhart dated February 3, 1945 in "State Department China/Japan/Korea lot files," RG 59, Box 5, held at the National Archives, Washington, D.C.
100. Ibid., summary of telephone conversation between Pearson and Vincent dated February 5, 1945.
101. Diary entry dated April 5, 1945, in Box 8, file 2, K. P. Chen Papers.
102. Cable to Weng Wenhao received May 5, 1945. See Ministry of Economics Archives 18–22–259 in *AS*. Article 4 of the contract read, in part, "each (repayment) note shall be unconditionally guaranteed, as to both principal and interest, by the Bank of China..." and "the government of the Republic of China...guaranteed the punctual payment...(and) that United States dollar exchange will be made available..."
103. Weng's memo dated May 22, 1945 in the Archives of the National Government (headquarters) 1120.1032.1 held at *AH*.
104. Generalissimo's office order dated May 29, 1945 in ibid.
105. Li Zhichuan and Chen Xinwen, *Hou Debang* (Tianjin: Nankai daxue chubanshe, 1986), 205.

106. Diary entry dated April 5, 1945 in K. P. Chen papers; and letter from Fan Xudong to Yan Youpu dated April 17, 1945, reprinted in *Haiwang* 20.26 (1948), 403.
107. On the possibility of Franklin Ho's effort on behalf of Fan, see Yu Xiaoqiu's manuscript memoirs, Q264-1-1216 held at *SM*. For Fan's criticisms of Weng, see Shou Chongyi, "Huiyi Fan Xudong xiansheng wan'nian ersanshi," [Remembering Fan Xudong's Last Years] *Gongshang jingji shiliao congkan* 2 (1983), 31–2.
108. Qian Changzhao (1998), 99.
109. Diary entries dated September 2–6, 1945 in Box 10, Zhang Jia'ao papers held at the Hoover Institution.
110. Joint memo to the Department of State from the China-American Council of Commerce and Industry and the National Foreign Trade Council dated December 20, 1946, Box 8, file 5, K.P. Chen papers; and George Wei (1997), 40–6.
111. Yun Zhen, "Ziyuan weiyuanhui de jishu yinjin gongzuo," [Technology transfer under the National Resources Commission] in Quanguo zhengxie wenshi ziliao yanjiu weiyuanhui gongshang jingjizu, ed., *Huiyi guomindang zhengfu ziyuan weiyuanhui* [Remembering the Nationalist Government's National Resources Commission] (Beijing: Zhongguo wenshi chubanshe, 1988), 170; and Zheng Youkui et al., *Jiuzhongguo ziyuan weiyuanhui* [National Resources Commission in Old China] (Shanghai: Shanghai shehui kexueyuan chubanshe, 1991), 193–4. Song allegedly also offered to authorize the guarantee provided that he, as chairman of the Bank of China, be made chairman of Yongli. See Zhang Tongyi (1987), 141–2. However, it was Kong Xiangxi who held the bank's chairmanship from 1944 to 1948. See Zhongguo yinhang hangshi bianji weiyuanhui comp., *Zhongguo yinhang hangshi* [A History of the Bank of China] (Beijing: Zhongguo jinrong chubanshe, 1995), 887–8.
112. Executive Yuan directive dated August 4, 1945, and decision to table dated October 9, 1945 in Ministry of Economics Archives 18–22–260 held at *AS*.
113. Fan and Zhou reportedly fell out in 1938 over the future of resistance. With Zhou implicated as a collaborator in occupied Shanghai, Fan, in his capacity as a member of Jincheng's supervising board, led an attempt in Chongqing to replace Zhou as chairman of the bank. See, respectively, Huang Hanrui, "Yi Sun Xuewu xiansheng," [Remembering Mr. Sun Xuewu] in Tan Jie et al., eds. (1997), 114; and diary entry dated March 21, 1942, in Box 18, Zhang Jia'ao papers held at the Hoover Institution.
114. Diary entry dated July 11, 1945 in Box 18, Zhang Jia'ao papers; and "A word to the wise" dated December 7, 1945 in Box 8, file 2, K.P. Chen papers.
115. *DGB* (Chongqing), October 5, 1945.
116. On Song and Zhou during the war, see Shen Zuwei and Du Xuncheng, eds. (1996), 266–91.
117. "Report to shareholders, 1937–1942," January 1943, Jiuda Salt Industries.
118. "Kangzhan wen'an" [Files on the War of Resistance] 302, *juan* 179-2 held at *AH*.

119. *ZGJDYWSZL* IV: 107–19 and Li Jianchang, *Guanliao ziben yu yanye* [Bureaucratic Capital and the Salt Industry] (Beijing: Sanlian shudian, 1963), 91–122.
120. Company papers 18–23–0-74–37–000–002 held at *AS*.
121. Order issued on December 22, 1945, facilitated by Li Zhuchen's old and new Hunanese friends in charge of the disposal of enemy properties in North China. See Sun Xihua, "Wo yu Li Zhuchen xiansheng jiaowangji," [My Association with Li Zhuchen] *Xiangxi wenshi ziliao* [Historical Material of Western Hunan] 25–26 (1992), 79–83. Miao Qiujie's confidential memo to Song Ziwen dated July 15, 1946 in Ministry of Economics Archives 18–36a, held at *AS*.
122. *Zhongyang ribao* [Central daily], June 18, 1943.
123. Under the "joint venture," Jiuda preexisting assets were valued at 448,000 *luanbi*; and Kahoku engyō invested another 1.25 million *luanbi*. See Uchida Keizō, "Kahoku engyō kabushiki kaisha hikitsugi chiyou shiyo" [Balance Report of the North China Salt Co., Ltd.] in Archives of Hebei Pingjinqu diwei chanye chuliju held at *TM*.
124. YL 301005, Hou Debang's letter to Zhang Qun dated October 7, 1947; and Li's petition dated December 6, 1947.
125. Ren Zhiyuan, "Jiuda de wangji yu xin shiming," [Jiuda's Past Accomplishments and New Mission] *Haiwang* 20.6 (1947), 86–7.
126. Fu Manyun's letters to Yu Xiaoqiu dated December 22, 1947 and January 8, 1948 in YL 310007–301010.
127. See YL 300969, 301007–11. Hence the company was not coopted by the Nationalist state as implied in Qi Zhilu, "Kangzhan shiqi gongkuang neiqian yu guanliao ziben de lueduo," [Relocation of Industries to the Interior and Their Appropriation by Bureaucratic Capital] *Gongshang jingji shiliao congkan* 2 (1983), 95.

Postscript

1. *DGB* (Chongqing) October 22, 1945.
2. Entry dated January 3, 1940, in Cao Boyen ed., *Hu Shi riji* [Diary of Hu Shi], 10 vols. (Taibei: Lianjing chubanshe, 2004), VIII: 4.
3. Entry dated January 11, 1941 in ibid., VIII: 90, and Fan's letter to Hu Shi dated September 13, 1942, in Hu Shi Archives held at the Special Collections, Institute of Modern History, Chinese Academy of Social Sciences, Beijing.
4. Yang Rengen, "Gongye fazhan yu xianshi zhengzhi," [Industrial Development and Political Reality] *Haiwang* 20: 24 (1948), 371–2; and Sun Guoda (1991), 146–51.
5. Wu (Sun Xuewu), "Duiyu gongchang shiye de xingzhi lingyi kanfa," [Another Perspective on the Nature of Industrial Enterprise] *Haiwang* 6.6 (1933), 81–82.
6. For the performance and fate of Jiuda after 1949, see Appendix 4b–e and my "Janus-faced capitalist: Li Zhuchen and the Jiuda Salt Refinery, 1949–1953," in Sherman Cochran, ed., *The Capitalist Dilemma* (Ithaca, NY: Cornell East Asia Series, forthcoming 2014).

7. Richard Whitley and Peer Hull Kristensen, eds., *The Changing European Firm: Limits to Convergence* (London: Routlege, 1996), passim; Vipin Gupta and Wang Jifu, "The Transvergence Proposition under Globalization: Looking Beyond Convergence, Divergence and Crossvergence," *Multinational Business Review* 12.2 (2004), 37–58; and the work of David A. Ralston et al.
8. Minutes of meeting, February 17, 1929, prepared by H. G. Allen, chairman, Imperial Chemical Industries (China) in the Archives of Brunner, Mond and Co., W45/3/73 held at Cheshire Local Archives.
9. On the growing literature on the Nationalist government's industrial policy and planned (controlled) economy, inspired by liberal or fascist economic theories, see Xu Jiansheng (2006), ch. 6; and Zanasi (2006).
10. Report on Fan dated March, 1941 in "Personnel Files," Generalissimo's Office held at Academia Historica, Taibei.
11. Zhang Zhizhong, "Wo suo zhidao de aiguo shiyejia Fan Xudong," [The Fan Xudong that I knew] *Hunan wenshi ziliao xuanji* [Selected Historical Materials on Hunan] 17 (1983), 27, 32; Qian Changzhao, (1988), 1; and Xie Yi, *Guomin zhengfu Ziyuan weiyuanhui yanjiu* [Studies on the Nationalist government's National Resources Commission] (Beijing: Shehui kexue wenxian chubanshe, 2005), 213.
12. James Reardon-Anderson thus erred in suggesting that Yongli's Nanjing plant was financed by a 5.5 million yuan state loan. See Reardon-Anderson (1991), 265.
13. Fan's letter to Fu Bingzhi dated May 31, 1941, in *YL* 300774.
14. The literature on capitalisms is too big to cite in full here. See, for examples, Steward R. Clegg and S. Gordon Redding, eds., *Capitalism in Contrasting Cultures* (New York: Walter de Gruyter, 1990); Richard Whitley (1999); Peter Hall and David Soskice, eds., *Varieties of Capitalism* (Oxford: Oxford University Press, 2001); and William Baumol, Robert E. Litan, and Carl J. Shramm, *Good Capitalism, Bad Capitalism* (New Haven, CT: Yale University Press, 2007).
15. See, for example, S. Gordon Redding (1990), and id., "What is Chinese about Chinese Family Business," in Henry Wai-chung Yeung and Kris Olds, eds., *Globalization of Chinese Firms* (London: Macmillan, 2000), 31–54. On corruption, see Huang Yasheng, *Capitalism with Chinese Characteristics* (Cambridge: Cambridge University Press, 2008), 236.
16. Matthias Schramm and Markus Taube, "Private Ordering of Corrupt Transactions," in Johann Graf Lambsdorff, Markus Taube, and Matthias Schramm, eds., *The New Institutional Economics of Corruption* (London: Routledge, 2005), 181–97. On the theory of progression of capitalisms, see Chu Yin-wah ed., *Chinese Capitalisms* (New York: Palgrave Macmillan, 2010), 14; and Victor Nee and S. Opper, "On Politicized Capitalism," in Victor Nee and R. Swedburg, eds. (2007), 93–127. For a review of the revisionist school on corruption, see Sun Yan, *Corruption and Market in Contemporary China* (Ithaca, NY: Cornell University Press, 2004), 13–17.
17. Rey Chow, "On Chineseness as a Theoretical Problem," *boundary 2* 25.3 (1998), 1–24; and Souchou Yao, *Confucian Capitalism: Discourse, Practice and the Myth of Chinese Enterprise* (London: RoutledgeCurzon, 2002), passim.

18. On grobalization, see George Ritzer, *The McDonaldization of Society* (20th anniversary ed., Los Angeles, CA: Sage Publications, 2013), 168. See also David Schak, "Chineseness and Chinese Capitalism in East and Southeast Asia," in Leo Douw, Cen Huang, and David Ip, eds., *Rethinking Chinese Transnational Enterprises* (Richmond, Surrey: Curzon, 2001), 83–101; and Edmund Terence Gomez and Hsin-huang Michael Hsiao, eds., *Chinese Enterprise, Transnationalism, and Identity* (London: RoutledgeCurzon, 2004), passim.

Glossary

aiguo juan	爱国捐
Anqing	安庆
baoxiao	包销
Bian Baimei	卞白眉
Cai E	蔡锷
Caizheng weiyuanhui	财政委员会
Cao Kun	曹锟
chan bu jian yun	产不兼运
Changlu	长芦
Chen Changheng	陈长蘅
Chen Guangfu	陈光甫
Chen Guangyuan	陈光远
Chen Jingmin	陈敬民
Chen Tianji	陈天骥
chengben jianqing fei	成本减轻费
Dalu xinwenshe	大陆新闻社
Dapu	大浦
Datong	大同
daixiao	代销
dingli weichi	鼎力维持
fa zhi	法治
Fan Gaoping	范高平
Fan Hongtao	范洪涛
Fan Xudong	范旭东
Fan Yuanlian	范源濂
Fang Jilin	方积琳
Fu Dingyi	符定一
Gonghengmao	公恒茂
Gong Xinzhan	龚心湛
Gongyi	公义
guanxiwang	关系网

guohuo	国货
Han Yongqing	韩永清
Hengfengtang	恒丰堂
Hankou	汉口
Hedong	河东
Hou Debang	侯德榜
huagong xiandao gongzai zhonghua	化工先导功在中华
Huanghai huaxue gongye yanjiushe	黄海化学工业研究社
Huangjia zhentanshe	皇家侦探社
Huang Wenzhi	黄文治
jiezhi ziben	节制资本
Jincheng yinhang	金城银行
Jing Benbai	景本白
Jingyan gongchang lianhe banshichu	精盐工厂联合办事处
Jingyan teshui	精盐特税
Jingyan tongze	精盐通则
Jingyan zonghui	精盐总会
Jing Youyan	荆有岩
jinyu	斤余
jiuchang zhengshui	就场征税
jiuchang zhuanmai	就场专卖
Jiuda jingyan gongsi	久大精盐公司
Jiuguo ribao	救国日报
Jiuhe	久和
Jiujiang	九江
juling zhizhang	巨灵之掌
Kong Xiangxi	孔祥熙
Lequn	乐群
Li Binshi	李宾士
li jingyan	粒精盐
Li Shiwei	李士伟
Li Sihao	李思浩
Li Wen	李雯
Li Yuanhong	黎元洪
Li Zhuchen	李烛尘
Lianghuai	两淮
Liang Qichao	梁启超
Liang Shiyi	梁士诒
Liangzhe	两浙
lingxing goushi	零星购食
Liu Kuiyi	刘揆一
lixing zhiyong	力行致用

luhao	卤耗
madainei baozhi fenyan	麻袋内包纸粉盐
Ma Taijun	马泰钧
Ma Yinchu	马寅初
menlu	门路
Miao Qiujie	缪秋杰
Minsheng	民生
minshi	民食
neidi	内地
Pan Fu	潘复
Pingzhengyuan	平政院
Qian Changzhao	钱昌照
Qian Junda	钱隽达
qingyi	情谊
Quanhua huaxue gongye Gongshi	全华化学工业公司
ren zhi	人治
shangbu	商埠
shefa tongrong hezhun	设法通融核准
shili	势力
Shi Shangkuan	史尚宽
shitong	十同
Song Ziwen	宋子文
Suwushu	苏五属
Sun Jia'nai	孙家鼐
Sun Xuewu	孙学梧
Tanggu	塘沽
Tang Hansan	唐汉三
Tang Shouqian	汤寿潜
tongnian	同年
tongshang kou'an	通商口岸
Tongyi	通益
tongzhi jingji	统制经济
Wang Daxie	汪大燮
weichi	维持
Wei Tingsheng	卫挺生
Weng Wenhao	翁文灏
Wuhe	五和
Wu Jinyin	吴金印
wuwei jingzheng	无谓竞争
wuyuan	五缘
Xiao Yaonan	肖燿南
Xinda	新达

Xiong Xiling	熊希龄
Xu Kan	徐堪
Xu Zhaozhou	许肇周
Yan Xiu	严修
yangma baoxiao	洋码报效
Yang Du	杨度
yanhutu	盐糊涂
Yanzheng taolunhui	盐政讨论会
Yanzheng zazhi	盐政杂志
yangsi	洋私
yanzhengfei	验证费
Yao Yongbai	姚詠白
Ye Gengxu	叶耕绪
Ye Kuichu	叶揆初
Yihexiang	乙和祥
Yingcheng	应城
yinpiao	引票
Yongli zhijian gongsi	永利制碱公司
Yongyu	永裕
Yu Xiaoqiu	余啸秋
yuan'an banli	原案办理
yunxiao gongyue	运销公约
zai	载
Zeng Yangfeng	曾养丰
Zhang Guogan	张国淦
Zhang Jian	张謇
Zhang Hu	张弧
Zhang Qinji	章勤济
Zhang Qun	张群
zhibao fenyan	纸包粉盐
Zhonghua wenhua jijinhui	中华文化基金会
Zhong Lijian	钟履坚
Zhong Shiming	钟世铭
Zhou Qingyun	周庆云
Zhou Xuexi	周学熙
Zhou Zuomin	周作民
zikai shangbu	自开商埠
Zou Bingwen	邹秉文
Zou Lin	邹琳
Zhu Tingqi	朱庭祺
zujie	租界

Selected Bibliography

Adshead, S. A. M. *The Modernization of the Chinese Salt Administration.* Cambridge, MA: Harvard University Press, 1970.
Adshead, S. A. M. *Salt and Civilization.* New York: St. Martin's Press, 1992.
Barthélemy, Jérôme. "Agency and Institutional Influences on Franchising Decisions." *Journal of Business Venturing* 26.1 (2011), 93–103.
Bergère, Marie-Claire. *The Golden Age of the Chinese Bourgeoisie.* Transl. Janet Lloyd. Cambridge: Cambridge University Press, 1989.
Bian Jintao. "Jiuda jingyan gongchang fangwenji" [A Visit to the Jiuda Salt Refinery]. *Wenhua jianshe* [Cultural Construction] 2.6 (1936), 81–86.
Bian, Morris L. *The Making of the State Enterprise System in Modern China.* Cambridge, MA: Harvard University Press, 2005.
Biggart, Nicole W. and Gary G. Hamilton. "On the Limits of a Firm-based Theory to Explain Networks: Western Bias of Neoclassical Economics." In *Networks and Organizations,* ed. Nitin Nohria and Robert G. Eccles, 471–490. Cambridge, MA: Harvard Business School, 1992.
Bräutigam, Deborah, Odd-Helge Fjeldstad and Mick Moore, eds. *Taxation and State-Building in Developing Countries.* Cambridge: Cambridge University Press, 2008.
Brodie, Patrick. *Crescent Over Cathay.* London: Oxford University Press, 1990.
Brook, Timothy. *Collaboration: Japanese Agents and Local Elites in Wartime China.* Cambridge, MA: Harvard University Press, 2005.
Brubaker, Rogers. "In the Name of the Nation: Reflections on Nationalism and Patriotism." In *The Many Faces of Patriotism,* ed. Philip Abbott, 37–52. Lanham: Rowman and Littlefield, 2007.
Bush, Richard C. *The Politics of Cotton Textiles in Kuomintang China.* New York: Garland Press, 1982.
Caizhengbu Yanwushu, comp. *Yanwu taolunhui huiyi huibian* [Proceedings of the Conference on Salt Affairs]. Nanjing: Ministry of Finance, 1928.
Caizhengbu Yanwushu. *Yanwu nianjian* [Annual Report on Salt Affairs]. Nanjing: Caizhengbu Yanwushu, 1930.
Caizhengbu Yanwushu. *Min'guo ershinerian Yanwu jihesuo nianbao* [Annual Report of the Revenue Inspectorate for 1933]. Nanjing: Caizhengbu, 1934.
Caizhengbu Yanwu zongju, comp. *Yanwu faling huibian* [Compendium of Salt Laws and Regulations]. Chongqing: Caizhengbu Yanwu zongju, 1942.

Cao Boyen, ed. *Hu Shi riji quanji* [Complete Diary of Hu Shi]. 10 vols. Taibei: Lianjing chubanshe, 2004.

Chan, Wellington. "Tradition and Change in the Chinese Business Enterprise." In *Chinese Business History*, ed. Robert Gardella, Jane K. Leonard, and Andrea McElderry, 127–144. New York: M.E. Sharpe, 1998.

Chandler, Alfred. *Strategy and Structure*. Cambridge, MA: MIT Press, 1962.

Chen Canglai. *Zhongguo yanye* [China's Salt Industry]. Shanghai: Shangwu yinshuguan, 1928.

Chen Guyuan. *Shuangqingshi yuwen cungao xuanlu* [Selected Left-over Writings from *Shuangqingshi*]. Taibei: N.p., 1965.

Chen Xinwen. *Fan Xudong*. N.p., 2002.

Chen Zhongping. *Modern China's Network Revolution: Chambers of Commerce and Sociopolitical Change in the Early Twentieth Century*. Stanford, CA: Stanford University Press, 2011.

Chien, Cecilia Lee-fang. *Salt and State*. Ann Arbor, MI: Center of Chinese Studies, University of Michigan, 2004.

Chu Yin-wah ed. *Chinese Capitalisms*. New York: Palgrave Macmillan, 2010.

Clegg, S. R. and S. G. Redding, eds. *Capitalism in Contrasting Culture*. Berlin: Walter de Gruyter, 1990.

Chow, Rey. "On Chineseness as a Theoretical Problem." *boundary 2* 25.3 (1998), 1–24.

Coase, Ronald H. "The Nature of the Firm." *Economica*, New Series 4 (1937), 386–405.

Coase, Ronald H. "The Nature of the Firm: Origin." *Journal of Law, Economics, & Organization* 4.1 (1988), 3–17.

Coble, Parks. *The Shanghai Capitalists and the Nationalist Government*. Cambridge, MA: Harvard University Press, 1980.

Coble, Parks. *Chinese Capitalists in Japan's New Order*. Berkeley, CA: University of California Press, 2003.

Cochran, Sherman. *Big Business in China*. Cambridge, MA: Harvard University Press, 1980.

Cochran, Sherman. *Encountering Chinese Networks*. Berkeley, CA: University of California Press, 2000.

Cochran, Sherman and Andrew Hsieh. *The Lius of Shanghai*. Cambridge, MA: Harvard University Press, 2013.

Davis, Clarence C. "Financing Imperialism: British and American Bankers as Vectors of Imperial Expansion in China, 1908–1920." *Business History Review* 56.2 (1982): 236–264.

Dayer, Roberta Allbert. *Bankers and Diplomats in China*. London: Frank Cass, 1981.

Dayer, Roberta Allbert. *Finance and Empire*. New York: St. Martin's Press, 1988.

Ding Changqing. "Guomindang Nanjing zhengfu yu jiuyanshang" [The Nationalist Nanjing Government and the Traditional Salt Merchants]. *YYSYJ* 1 (1988), 41–48.

Ding Changqing, ed. *Minguo yanwu shigao* [A Draft History on Republican Salt Affairs]. Beijing: Renmin chubanshe, 1990.

Ding Changqing and Tang Renyue, eds. *Zhongguo yanyeshi* [A History of China's Salt Industry]. 3 vols. Beijing: Renmin chubanshe, 1997.
Dirlik, Arif. "Critical Reflections on 'Chinese Capitalism' as Paradigm." *Identities* 3.3 (1997), 303–330.
Dong Zhengping. *Kangzhan shiqi Guomin zhengfu yanwu zhengce yanjiu* [Salt Policy of the Nationalist Government during the War of Resistance]. Ji'nan: Qilu shushe, 2004.
Douw, Leo, Cen Huang and David Ip, eds. *Rethinking Chinese Transnational Enterprises*. Richmond, Surrey: Curzon, 2001.
Du Xuncheng. *Minzu ziben zhuyi yu jiu Zhongguo zhengfu* [National Bourgeois Capitalism and Old China's Government]. Shanghai: Shanghai shehui kexueyuan chubanshe, 1991.
Eastman, Lloyd E. et al. *The Nationalist Era in China, 1927–1949*. Cambridge: Cambridge University Press, 1991.
Fang Qingqiu, et al. *Minguo shehui jingjishi* [Social and Economic History of the Republican Period]. Beijing: Zhongguo jingji chubanshe, 1991.
Faure, David. *China and Capitalism*. Hong Kong: Hong Kong University Press, 2006.
Fruin, Mark ed. *Networks, Markets, and the Pacific Rim*. New York: Oxford University Press, 1998.
Gale, Esson M. *Salt for the Dragon*. East Lansing, MI: Michigan State College Press, 1953.
Gargiulo, Martin and Mario Benassi. "The Dark Sides of Social Capital." In *Corporate Social Capital and Liability* ed. Th. A. Roger, J. Leenders, and Shaul M. Gabby, 298–322. Norwell, MA: Kluwer Academic Publishing, 1999.
Granovetter, Mark. "The Strength of Weak Ties." *American Journal of Sociology* 78 (1973), 1360–1380.
Granovetter, Mark. "Economic Action and Social Structure: The Problem of Embeddedness." *American Journal of Sociology* 91.3 (1985), 481–510.
Granovetter, Mark. "Coase Revisited: Business Groups in the Modern Economy." *Industrial and Corporate Change* 4 (1995), 93–130.
Greenfeld, Liah. *The Spirit of Capitalism*. Cambridge, MA: Harvard University Press, 2001.
Grossman, Peter Z. ed. *How Cartels Endure and How They Fail*. Cheltenham, UK: Edward Elgar Publishing, 2004.
Guojia shuiwu zongju, comp. *Zhonghua minguo gongshang shuishou shi—yanshui juan* [A History of Republican Tax—Gabelle]. Beijing: Zhongguo caizheng jingji chubanshe, 1999.
Guo Zhengzhong, Ding Changqing, and Tang Renyue, eds. *Zhongguo yanyeshi* [A History of Salt in China]. 3 vols. Beijing: Renmin chubanshe, 1997.
Gupta, Vipin and Wang Jifu. "The Transvergence Proposition under Globalization: Looking Beyond Convergence, Divergence and Crossvergence," *Multinational Business Review* 12.2 (2004), 37–58.
He Shimi. *Kangzhan shiqi zhuanmai shiliao* [Historical Materials of State Monopolies during the War of Resistance]. Taibei: Academia Historica, 1992.

Hite, Julie M. "Patterns of Multidimensionality among Embedded Network Ties: A Typology of Relational Embeddedness in Emerging Entrepreneurial Firms." *Strategic Organization* 1.1 (2003), 9–49.
Hsiung, James C. and Steven I. Levine, eds. *China's Bitter Victory*. New York: M.E. Sharpe, 1992.
Huagong xiandao Fan Xudong biansheju comp. *Huagong xiandao Fan Xudong* [Fan Xudong: a Pioneer of Chemical Industry]. Beijing: Zhongguo wenshi chubanshe, 1987.
Huang, Philip C. C. *Liang Ch'i-chao and Modern Chinese Liberalism*. Seattle: University of Washington, 1972.
Huang Yasheng. *Capitalism with Chinese Characteristics*. Cambridge: Cambridge University Press, 2008.
Huang Yiping, et al. *Beiyang zhengfu shiqi jingji* [The Chinese Economy during the Northern Warlord Period]. Shanghai: Shanghai shehui kexueyuan, 1995.
Jiang Fenglan. *Guomin zhengfu shiqi de yanzheng shiliao* [Historical Materials on Salt Administration under the Nationalist Government]. Taibei: Academia Historica, 1993.
Jiang Liangqin. *Nanjing guomin zhengfu neizhai wenti yanjiu* [Nanjing Government's Problem of Domestic Debt]. Nanjing: Nanjing daxue chubanshe, 2003.
Keister, Lisa A. *Chinese Business Groups*. Oxford: Oxford University Press, 2000.
Kirby, William C. "China Unincorporated: Company Law and Business Enterprise in Twentieth-Century China." *Journal of Asian Studies* 54.1 (February 1995), 43–63.
Kishi Toshihiko. "Yongli kagaku kōgyō konsu to Fan Xudong." In Soda Saburō ed. *Chūgoku kindaika katei no shidōsha tachi* [Leaders in China's Modernization Process]. Tokyo: Tōhō shoten, 1997.
Kōgyō kagaku kai Manshū shibu, ed. *Manshū no shigen to kagaku kōgyō* [Resources and Chemical Industries in Manchuria]. Tokyo: Maruzen kabushiki kaisha, 1937.
Köll, Elisabeth. *From Cotton Mill to Business Empire* Cambridge, MA: Harvard University Asia Center, 2003.
Krippner, Greta, Mark Granovetter, Fred Block, et al. "Polanyi Symposium: A Conversation on Embeddedness." *Socio-Economic Review* 2 (2004), 109–135.
Laird, Pamela Walker. *Pull*. Cambridge, MA: Harvard University Press, 2006.
Lamoreaux, Naomi. *Insider Lending: Banks, Personal Connections, and Economic Development in Industrial New England*. Cambridge: Cambridge University Press, 1994.
Lamoreaux, Naomi R., Daniel M.G. Raff, and Peter Temin. "Beyond Markets and Hierarchies: Toward a New Synthesis of American Business History." *NBER Working Paper No. 9029* (July, 2002).
Lazonick, William. *Business Organization and the Myth of the Market Economy*. Cambridge: Cambridge University Press, 1991.
Levenstein, Margaret C. and Valerie Y. Suslow. "What Determines Cartel Success?" *Journal of Economic Literature* XLIV (2006), 43–95.
Li Han. *Miao Qiujie yu Minguo yanwu* [Miao Qiujie and Salt during the Republican Period]. Beijing: Zhongguo kexue jishu chubanshe, 1990.

Li Jianying. "Bumen chuangxin tixi jianshezhong de Yongli wushi" [Sectoral Innovations and Institutional Building: the Yongli Model]. PhD Dissertation, Nankai University, 2009.

Li Yu. *Beiyang zhengfu shiqi qiye zhidu jiegou shilun* [On the Structure of the Corporate System during the Northern Warlord Period]. Beijing: Shehui kexue wenxian chubanshe, 2007.

Li Zhichuan and Chen Xinwen. *Hou Debang*. Tianjin: Nankai daxue chubanshe, 1986.

Liu Changshan. *Qingdai houqi zhi minguo chunian yanwu de bianqe* [Changes in Salt Administration from Late Qing to Early Republic]. Taibei: Wenshizhe chubanshe, 2006.

Liu Cunliang. "Zhongguo renmin zhi yanshui fudan" [The Salt Tax Burden Borne by the Chinese People]. *Zhongguo jingji* [China's Economy] 2.2 (1934), 1–40.

Lipartito, Kenneth and David B. Sicilia, eds. *Constructing Corporate America*. New York: Oxford University Press, 2005.

Liu Shoulin, Wan Renyuan, Wang Yumin, and Kong Qingtai, et al., eds. *Minguo zhiguan nianbiao* [Chronological Dictionary of Republican Officials]. Beijing: Zhonghua shuju, 1995.

Liu Wenbin. *Jindai Zhongguo qiyue guanli sixiang yu zhidu de yanbian* [The Thought and Institutional Transformation of Management in Modern China]. Taibei: Academia Historica, 2001.

Loveridge, Ray. "The Firm as Differentiator and Integrator of Networks." In David O. Faulkner and Mark de Rond, eds. *Cooperative Strategy: Economic, Business, and Organizational Issues*,135–169. Oxford: Oxford University Press, 2000.

Ma Yinchu. Caizhengxue yu Zhongguo caizheng [Fiscal Science and China's Public Finance]. 2 vols. Shanghai: Shangwu yinshuguan, 1948.

MacKenzie, Donald, Fabian Muniesa, and Lucia Siu, eds. *Do Economists Make Markets?* Princeton: Princeton University Press, 2007.

MacMurray, John V. A. comp. and ed. *Treaties and Agreements with and Concerning China*, 2 vols. New York: Oxford University Press, 1921.

McChesney, Fred. *Money for Nothing*. Cambridge, MA: Harvard University Press, 1997.

McLean, David. "British Banking and Government in China: the Foreign Office and the Hongkong and Shanghai Bank, 1895–1914." PhD dissertation, Cambridge University, 1973.

Motta, Massimo. *Competition Policy: Theory and Practice*. Cambridge: Cambridge University Press, 2004.

Nee, Victor and R. Swedburg, eds. *On Capitalism* Stanford, CA: Stanford University Press, 2007.

Pearson, Margaret M. *China's New Business Elite*. Berkeley, CA: University of California Press, 1997.

Peng Zeyi and Wang Renyuan, eds., *Zhongguo yanyeshi guoji xueshu taolunhui lunwenji* [Proceedings of the International Symposium on the History of China's Salt Industry]. Chengdu: Sichuan renmin chubanshe, 1991.

Perkins, Dwight ed. *China's Modern Economy in Historical Perspective.* Stanford, CA: Stanford University Press, 1975.
Pong, David ed. *Resisting Japan.* Norwalk, CT: Eastbridge, 2008.
Powell, Walter W. "Neither Market nor Hierarchy: Network Forms of Organization." *Research in Organizational Behavior* 11 (1990), 295–336.
Primoratz, Igor, ed. *Patriotism.* Amherst, New York: Humanity Books, 2002.
Qian Changzhao. *Qian Changzhao huiyilu* [Memoirs of Qian Changzhao]. Beijing: Zhongguo wenshi ziliao chubanshe, 1998.
Qingdaoshi shehuiju comp. *Qingdaoshi yanye diaocha* [Survey of Qingdao's Salt Industry]. Qingdao: Qingdaoshi shehuiju, 1933.
Quanguo jingji huiyi mishuchu, comp. *Quanguo jingji huiyi zhuankan* [Special Collection of Proposals from the National Economic Conference]. N.p, 1928.
Ralston, David A. "The Crossvergence Perspective: Reflections and Projections." *Journal of International Business Studies* 39.1 (2008), 27–40.
Rawski, Thomas G. "China's Republican Economy: An Introduction." Discussion paper, University of Toronto-York University Joint Centre on Modern East Asia, 1978.
Rawski, Thomas G. *Economic Growth in Prewar China.* Berkeley, CA: University of California Press, 1989.
Reardon-Anderson, James. *The Study of Change: Chemistry in China, 1840–1949.* Cambridge: Cambridge University Press, 1991.
Redding, S. Gordon. *The Spirit of Chinese Capitalism.* Berlin: Walter de Gruyter, 1990.
"Refined-salt Industry in China." *Chinese Economic Journal* 13.1 (1933): 65–71.
Richter, Frank-Jurgen. *Business Networks in Asia.* Westport, CT: Quorum Books, 1999.
Ritzer, George. *The McDonaldization of Society.* 20th anniversary ed., L.A.: Sage Publications, 2013.
Rotberg, Robert I. *Patterns of Social Capital.* Cambridge: Cambridge University Press, 2001.
Schumpeter, Joseph A. *Capitalism, Socialism and Democracy* London: Routledge, 1994.
Sheehan, Brett. *Trust in Troubled Times.* Cambridge, MA: Harvard University Press, 2003.
Shi Bolin. *Qifeng kuyu zhong de minguo jingji* [The Precarious Republican Economy]. Zhengzhou: Henan renmin chunbanshe, 1993.
Shoji Tsutomu, comp. *Nihon soda kogyōshi* [A History of Japan's Soda Ash Industry]. Tokyo: Soda sarashiko dogyōkai, 1938.
Stocking, George W. and Myron W. Watkins, et. al., *Cartels in Action: Case Studies in International Business Diplomacy.* New York: Twentieth Century Fund, 1946.
Strauss, Julia C. *Strong Institutions in Weak Polities.* Oxford: Clarendon Press, 1998.
Sun Guoda. *Minzu gongye daqianxi* [Mass Migration of the National Bourgeoisie]. Beijing: Zhongguo wenshi chubanshe, 1991.

Todeva, Emanuela. *Business Networks*. London: Routledge, 2006.
Uzzi, Brian. "The Sources and Consequences of Embeddedness for the Economic Performance of Organizations: The Network Effect." *American Sociological Review* 61.4 (1996), 674–698.
Van de Ven, Andrew H. and William F. Joyce, eds. *Perspectives on Organizational Design and Behavior*. New York: Wiley, 1981.
Wang Chuhui. *Zhongguo jindai qiyue zuzhi xingtai de bianqian* [Transformation of Business Organization in Modern China]. Tianjin: Renmin chubanshe, 2001.
Wang Fangzhong. "1927–1937 nianjian de Zhongguo yanwu yu yanfa gaige de liuchan" [China's Aborted Reform of Salt Administration and Law, 1927–1937]. In *Zhongguo yanyeshi guoji xueshu taolunhui lunwenji* [Proceedings of the International Symposium on the History of China's Salt Industry] ed. Peng Zeyi and Wang Renyuan, 159–181. Chengdu: Sichuan renmin chubanshe, 1991.
Watanabe Atushi. "Shimmatsu ni okeru ensei no chūō shūkenka seisaku ni tsuite." [On the Centralization Policy of Salt Administration in Late Qing.] In *Nakajima Satoshi sensei koki kinen ronshu* [Studies in Asian History Dedicated to Professor Satoshi Nakajima on His Seventieth Birthday] ed. Nakajima Satoshi senshi koki kinen jigyōkai, 2 vols., II: 657–680. Tokyo: Kyuko shoin, 1980.
Watson, John. "Modeling the Relationship between Networking and Firm Performance." *Journal of Business Venturing* 22.6 (2007), 852–874.
Wei, George C.X. *Sino-American Economic Relations, 1944–1949*. Westport, CT: Greenwood Press, 1997.
Wells, Wyatt. *Antitrust and the Formation of the Postwar World*. New York: Columbia University Press, 2002.
Whitley, Richard D. *Business Systems in East Asia: Firms, Markets, and Societies*. Beverly Hills: Sage, 1992.
Whitley, Richard D. *Divergent Capitalisms*. Oxford: Oxford University Press, 1999.
Whitley, Richard D. and Peer Hull Kristensen, eds. *The Changing European Firm: Limits to Convergence*. London: Routlege, 1996.
Williamson, Oliver E. *Markets and Hierarchies*. New York: The Free Press, 1975.
Williamson, Oliver E. *The Economic Institutions of Capitalism*. New York: The Free Press, 1985.
Williamson, Oliver E. "The New Institutional Economics: Taking Stock, Looking Ahead." *Journal of Economic Literature* XXXVII (Sept., 2000), 595–613.
Williamson, Oliver E. "The Theory of the Firm as Governance Structure: From Choice to Contract." *Journal of Economic Perspectives* 16.3 (2002), 171–195.
Windoff, Paul. *Corporate Networks in Europe and the U.S.* Oxford: Oxford University Press, 2002.
Wong Siu-lun. "The Chinese Family Firm: A Model." *British Journal of Sociology* 36:1 (1985): 58–72.

Wright, Tim. "The Spiritual Heritage of Chinese Capitalism: Recent Trends in the Historiography of Chinese Enterprise Mangement." In *Using the Past to Serve the Present: Historiography and Politics in Contemporary China*, ed. Jonathan Unger, 205–238. Armonk, NY: M.E. Sharpe, 1993.

Xu Jiansheng. *Minguo shiqi jingji zhengce de yanxi yu bianyi* [Inherited and Shifting Economic Policies of the Republican Period]. Fuzhou: Fujian renmin chubanshe, 2006.

Yan Sun. *Corruption and Market in Contemporary China*. Ithaca, NY: Cornell University Press, 2004.

Yang Xingqin. *Zhongguo zhanshi yanwu* [Salt Affairs in Wartime China]. N.p.: Guomin chubanshe, 1943.

Yao Shundong. "Nanjing guomin zhengfu chuqi shiyan lifa yu Zhongguo yanwu jindaihua" [Alimental Salt Legislation by the Early Nanjing National Government and Modernization of China's Salt Administration]. *YYSYJ* 1 (2007), 19–26.

Yao Souchou. *Confucian Capitalism: Discourse, Practice and the Myth of Chinese Enterprise*. London: RoutledgeCurzon, 2002.

Yeh Wen-hsin ed. *Becoming Chinese*. Berkeley, CA: University of California Press, 2000.

Young, Arthur N. *China's Nation-Building Effort*. Palo Alto, CA: Hoover Institution Press, 1971.

Zaiseibu Zeimushi enmuka, comp. *Enmu shiryō* [Materials on Salt]. Shinkyō: Zaiseibu, 1935.

Zanasi, Margherita. *Saving the Nation*. Chicago: University of Chicago Press, 2006.

Zelin, M. *The Merchants of Zigong*. New York: Columbia University Press, 2006.

Zeng Yangfeng. *Zhongguo yanzhengshi* [A History of Salt Administration in China]. Repr. Ed., Taibei: Shangwu yinshuguan, 1978.

Zhang Hongxiang. *Jindai Zhongguo tongshang kou'an yu zujie* [Treaty Ports and Foreign Concessions in Modern China]. Tianjin: Renmin chubanshe, 1993.

Zhang Lijie. "Tanxi Nanjing guomin zhengfu weineng shishi xinyanfa de yuanyin" [Why Did the Nanjing Government Fail to Implement the New Gabelle Law]. *YYSYJ* 3 (2007), 10–20.

Zhang Lijie. *Nanjing guomin zhengfu de yanzheng gaige yanjiu* [A Study of Salt Administration Reform under the Nanjing Government]. Beijing: Zhongguo shehui kexueyuan chubanshe, 2011.

Zhang Ning, Jennifer. "Vertical Integration, Business Diversification, and Firm Architecture: The Case of the China Egg Produce Company in Shanghai, 1923–1950." *Enterprise and Society* 6.3 (2005), 419–451.

Zhang Pengyuan. *Lixianpai yu Xinhai geming* [The Constitutionalists and the 1911 Revolution]. Taibei: Academia Sinica, 1969.

Zhang Tongyi. *Fan Xudong zhuan* [A Biography of Fan Xudong]. Changsha: Hunan renmin chubanshe, 1987.

Zhang Yufa. *Min'guo chu'nian de zhengdang* [Political Parties in Early Republican China]. 2nd ed., Taibei: Institute of Modern History, Academia Sinica, 2002.

Zhang Zhongmin. "Jindai Zhongguo gongsi zhidu yanjiu de huigu yu zhanwan" [A Review and Prospects of Studies on the Corporate System in Modern China]. In *Chugoku kigyoshi kenkyū no seika to katai* [Achievements and Issues in Chinese Business History], ed. Chugoku kigyoshi kenkyūkai, 47–61. Tokyo: Kyuko shoin, 2007.

Zukin, Sharon and Paul DiMaggio. *Structures of Capital*. Cambridge: Cambridge University Press, 1990.

Index

Anhui, 18, 70–1, 88, 192n. 37

Bian Baimei, 31, 72
bribery, 57, 59, 81, 96, 98, 138
 see also donation; gifts
Brunner, Mond, 2, 89

Cao Kun, 58, 68–70
cartel, 6, 77–8, 92
Changlu, 10, 20–1, 40–1, 44, 53, 71, 83, 102–4, 107
 free trade experiment, 18
 revenue farmers, 47, 62, 97
 see also Ding Naiyang; Jing You'an; Li Binshi
Changsha, 45
Chen Changheng, 97, 100, 109, 113
Chen Guangfu, 128, 129, 130, 131
Chen Guangyuan, 59, 68, 69
Chen Jingmin, 27, 29, 31, 32, 33, 34
Chen Tianji, 111–13
Cleveland, F.A., 19, 21, 101

daixiao, 39
Dane, Richard M., 13, 15–19, 35, 41
Datong, 79, 80
Ding Naiyang, 53–4
donation, 18, 103, 112, 113, 133
 compare bribery; gifts

Fan Gaoping, 82, 83, 85, 86, 92, 93
Fan Xudong
 death, 130

education, 25
entrepreneurship, 75, 90, 91, 126, 128
eulogized, 135, 137
family, 1, 25, 45
government service, 25, 109
network, 29, 30, 32, 39, 66, 93, 96, 97, 103, 109, 112, 119
post-war plan, 127–8
sons of Hunan, 87, 91
Fan Yuanlian, 25, 29, 33, 39, 46, 52, 185n. 69, 201n. 26
Fang Xiangeng, 63
Feng Guifen, 10
Finance Ministry, 12, 15, 99, 105, 107, 119, 126, 132
 see also republican governments; Fu Dingyi; Gong Xinzhan; Gu Yingfen; Kong Xiangxi; Li Sihao; Liang Qichao; Pan Fu; Song Ziwen; Xiong Xiling; Xu Gan; Zhang Hu; Zhang Shouyong; Zhong Shiming; Zhou Xuexi; Zou Lin
Fu Dingyi, 71, 81

Gale, Esson M., 9, 17, 22, 188n. 40, 190n. 64
gifts, 58, 59, 72, 113, 193n. 40, 194n. 59
 see also bribery; donation
Gu Yingfen, 96
Gong Xinzhan, 38, 62, 69

Han Yongqing, 33, 44
Hankou, 84, 87, 88, 90, 108
  see also Hubei
Hou Debang, 73, 91, 132
Hengfengtang, 1, 64, 66, 79, 120
Hu Shi, 97, 131, 135
Huang Wenzhi, 78, 79, 81, 93
Huanghai, 1, 64, 121–2
Hubei, 18, 56, 57, 58–9, 70, 71, 79, 81, 88, 116, 118, 121, 123
  Salt Bureau, 44–5, 82
  see also Hankou
Hunan, 10, 18, 67, 80, 110
  Salt Bureau, 81
  see also Changsha
Hunan shiwu xuetang, 25, 33
  see also Liang Qichao; Xiong Xiling

Jiang Jieshi, 2, 97, 101, 109, 127, 128, 135, 137
Jiangxi, 18, 45, 68–9
Jincheng yinhang, 30, 31, 66, 73–4, 76, 89, 107, 120
  see also Zhou Zoumin
Jing Benbai, 10, 17, 19, 20, 25, 27, 30–1, 33, 34, 35, 39, 40, 44–5, 59, 65, 70, 82, 91, 96, 103, 105
Jing Youyan, 102–4, 107
jiuchang zhengshui, 10
jiuchang zhuanmai, 11
Jiuda Salt Refinery Co., Ltd.
  bank network, 30–2, 39, 45, 66, 89, 107, 112, 120, 130
  capitalization, 30, 45, 59, 64
  Dapu, 117
  factionalism, 91
  franchising, 38–9, 44, 45, 47, 76, 85
  governance, 91–2
  management hierarchy, 79, 81, 92
  organization, 1, 64, 91
  partnerships, 44–6, 65, 66, 78, 120
  production, 34, 36, 39, 40, 46, 55–7, 65, 72, 90, 117, 118–19, 121, 132
  registration, 34–6, 49–51, 54, 99
  reregistration, 102–7, 126
  share price, 59
  shareholders network, 27–30, 32–4, 43, 44–5, 47, 54, 58, 59, 67, 68, 71, 73, 83, 93
  Tanggu, 34–5, 36, 64, 65, 116, 118, 122, 132
  Zhangjiaba, 117
  see also Datong; Hengfangtang; Taiyu; Xinda; Yongli; Yongyu
Jiujiang, 68, 83, 85, 86–7, 90
  see also Jiangxi

Li Binshi, 27, 28, 32, 33, 34, 67
Li Shiwei, 39
Li Sihao, 39, 41, 46, 52, 59, 185n. 70
Li Yuanhong, 32, 33, 43, 54, 57, 58, 69
Li Zhuchen, 73, 91, 116, 118, 132, 136, 137
Liang Qichao, 12, 15, 25, 27, 28, 29, 31, 32, 33, 34, 41, 46, 55
  see also network; Hunan shiwu xuetang
Liang Shiyi, 26, 32
Liu Kuiyi, 29, 32, 33, 67
lobbying, 18, 58, 67, 69, 70, 72, 101, 106, 120
Lockhart, O.C., 124, 129
Lu Zuofu, 117, 120, 128

Ma Taijun, 103, 106, 108, 111, 132
Miao Qiujie, 67, 87, 102, 112, 116, 119, 132

National Board of Salt Refineries, 77, 79, 80, 92, 99, 103, 105–6, 109, 110
network
  embeddedness, 26, 32
  heterarchy, 26
  limits, 39, 43, 72
  strong and weak ties, 5, 42, 59
  structural hole, 32, 72, 97

Index   233

transitivity, 31, 33, 36–7, 68–9
  see also shitong; wuyuan

Pan Fu, 52, 53, 54, 81
patriotic capitalism, 2, 6, 91, 128
Pingzhengyuan, 48, 49, 50–2, 53, 54, 55, 57

Qian Changzhao, 116, 127, 130
Qian Fangshi, 59
Qian Junda, 96

refined salt cartel
  governance, 82, 87
  pooled sales, 80, 83, 84, 85, 87
  price-fixing, 83, 84
  see also cartel; National Board of Salt Refineries
refined salt markets
  Anqing, 45, 84, 86–7, 90
  four Yangzi provinces, 18, 21, 46, 44, 90, 111
  Hankou, 44, 46, 48, 49, 75, 78, 81, 82, 83, 84, 85, 86, 88, 90, 108
  Jiujiang, 68, 75, 83, 85, 86, 87, 90
  Nanjing, 44, 48–9, 50, 98, 112
refineries, 63–4, 70, 83, 89
  Dinghe, 93, 106, 107, 203n. 58
  Fengtian, 67, 69, 79, 80, 84, 90
  Fuhai, 85, 90
  Hongyuan, 87, 90
  Lequn, 47, 61
  Liyuan, 79, 80, 82, 83, 86, 90
  Mingsheng, 63, 90, 107
  Tongda, 86, 102, 107
  Tongyi, 32, 62–3, 66–7, 69, 72, 78–9, 81, 82, 84–5, 86–7, 90, 93, 105, 106–7
  Wuhe, 63, 66, 85, 86, 101, 106
  Xianfu, 62
  Yongyu, 1, 64, 65–6, 72, 73, 75–6, 79, 80, 81, 85, 106, 107
  Yuhua, 86, 90

  see also cartel, National Board of Salt Refineries
Reorganization Loan, 3, 5, 9, 11, 14, 25, 35, 124
Salt Administration, 12–14, 16, 17, 18, 48, 51, 56, 57, 58, 89
  Beijing, 36, 39, 46, 47, 52, 58, 65, 66, 70
  Chongqing, 117, 120, 123, 124, 132
  Nanjing, 96, 100, 101, 104, 111
  see also Gong Xinzhan; Li Sihao; Liang Qichao; Ma Taijun; Miao Qiujie; Pan Fu; Qian Fangshi; Qian Junda; Xu Jingren; Zhang Hu; Zhong Shiming; Zhou Xuexi; Zou Lin; Zhu Tingqi
republican governments, 2, 3, 9, 10, 13, 43, 48, 80
  Beijing, 11, 12, 14, 15, 19–20, 22, 23, 44, 50, 55, 56, 65, 69, 71
  Nanjing (Nationalist), 4, 82, 95, 96, 97, 98, 101, 102, 103, 108, 109, 110, 113–14, 115, 116, 123, 124–5, 127, 128, 138
revenue farmers, 4, 10, 21, 38, 39, 49, 57, 58, 62, 68, 70, 79, 80, 95–7, 98, 102, 113
  Gonghengmao, 51, 52
  Gongmao, 66
  Gongyi, 46, 47
  Yihexiang, 48, 49, 50, 204n. 82
revenue farming, 9–10, 11, 18–19, 21, 38, 50, 96, 98, 100–1, 113, 124, 131
  Changlu, 18, 21, 46–7, 62, 97, 102
  Lianghuai, 10, 11, 18–19, 21, 47, 51, 58, 61, 79, 88, 96–7, 98, 109, 118
  Liangzhe, 63, 97, 101
  Shandong, 10, 21, 65–6
  Shanghai, 66

Salt Revenue Inspectorate, 3, 9, 12, 13, 15–17, 19, 20, 22, 44, 55, 56, 70–1, 72, 96, 97, 113
  and Jiuda, 35, 36, 37, 43, 44, 53, 54
  synarchy, 15, 17, 21, 59, 101
  see also Cleveland, F.A.; Dane, Richard M.; Gale, Esson M.; Lockhart, O.C.; Wolcott, R.D.
Salt tax, 3, 10, 11, 17, 19–21, 35, 36, 71–2, 113–14, 124–5
  collateral to the Reorganization Loan, 9, 35, 124
Shitong, 26, 31, 33
  see also network
Sichuan, 10, 21, 22, 102, 116, 118, 120, 121, 126
  salt merchants, 117, 118, 119, 120, 131
Sun Jia'nai, 62
Sun Xuewu, 64, 120, 137

Taiyu, 81
Tang Shouqian, 10, 34
tongshang kou'an, 12, 34, 47, 48–52
tongzhi jingji, 2, 123–4, 137

Wang Shoushan, 63
Wei Tingsheng, 100
Weng Wenhao, 117, 119, 126, 127, 129, 130
Wolcott, R.D., 17, 101, 124
Wu Dingchang, 29, 31, 33, 120
Wu Jinyin, 45, 68–9
Wuyuan, 26
  see also network; shitong

Xiao Yaonan, 58–9
Xinda, 44

Xinfu, 81
Xiong Xiling, 11, 20, 25, 29, 32, 33, 67
Xu Jingren, 96

Yang Du, 29, 32, 33, 39
yangsi, 51
Yanzheng taolunhui, 11, 34
Yao Yongbai, 29, 33, 83
Ye Kuichu, 29, 30, 32, 34, 59
yinpiao, 14
  see also Salt Administration
Yongli Soda Ash Co., Ltd., 64, 73
Yuan Shikai, 9, 11–12, 15, 18, 33, 39

Zeng Yangfeng, 103, 104, 105, 106, 132
Zhang Hu, 18, 32, 33, 37, 39, 41, 47, 52, 57, 59, 101
Zhang Jian, 10, 11, 26, 31, 33
Zhang Qinji, 28, 29, 33, 45, 46, 80
Zhang Qun, 117, 132
Zhang Shouyong, 108
Zhong Lijian, 103, 104, 105, 106, 108, 109–13, 118
Zhong Shiming, 55
Zhou Qingyun, 63, 66, 101
Zhou Xuexi, 14, 25, 31, 33, 36–7, 38–9
Zhou Zuomin, 27, 30, 31, 33, 72, 96, 97, 104, 105, 130, 212n. 113
  see also Jincheng yinhang
zikai shangbu, 38
Zou Bingwen, 112
Zou Lin, 97, 106, 108
Zhu Tingqi, 104, 109, 110, 112, 124

Printed and bound by CPI Group (UK) Ltd, Croydon, CR0 4YY